FM 3-22.31 (FM 23-31)
C1

Change 1
Field Manual
No. 3-22.31

Headquarters
Department of the Army
Washington, DC, 19 March 2007

40-mm Grenade Launcher, M203

1. Change FM 3-22.31, 13 February, 2003, as follows:

Remove old pages:	**Insert new pages:**
i through iv	I through iv
1-1 through 1-2	1-1 through 1-2
Chapter 2	Chapter 2
Chapter 3	Chapter 3
Chapter 4	Chapter 4
Chapter 5	Chapter 5
6-9 through 6-10	6-9 through 6-10
Glossary	Glossary
References	References
Index	Index
DA Form 2946-R	DA Form 2946-R

2. A star (*) marks new or changed material.

3. File this transmittal sheet in front of the publication.

DISTRIBUTION RESTRICTION: Approved for public release; distribution is unlimited.

By Order of the Secretary of the Army:

PETER J. SCHOOMAKER
General, United States Army
Chief of Staff

Official:

Joyce E. Morrow

JOYCE E. MORROW
Administrative Assistant to the
Secretary of the Army
0705702

DISTRIBUTION:
Regular Army, USAR, and ARNG: To be distributed in accordance with initial distribution number 115893, requirements for FM 3-22.31.

This page intentionally left blank.

FIELD MANUAL
NO. 3-22.31 (23-31)

HEADQUARTERS
DEPARTMENT OF THE ARMY
WASHINGTON, D.C., 19 March 2007

40-MM GRENADE LAUNCHER, M203

CONTENTS

DISTRIBUTION RESTRICTION: Approved for public release; distribution is unlimited.

*This publication supersedes FM 23-31, 20 September 1994.

***DA Form 2946-R**

PREFACE

This publication provides technical information on and training and combat techniques for the M203 grenade launcher. Intended users include leaders and designated grenadiers, who will use this information to successfully integrate the M203 into their combat operations. This manual discusses gunnery training and train-the-trainer, and it includes an appendix on the 40-mm grenade launcher, M79.

*This publication applies to the Regular Army, the Army National Guard (ARNG)/Army National Guard of the United States (ARNGUS), and the United States Army Reserve (USAR) unless otherwise stated.

*The tactical positions shown in this manual were drawn to enhance the reader's understanding of related subject material and are *not* tactically correct. Also, to avoid hiding important graphic information, uniforms may be shown plain rather than camouflaged.

*This publication prescribes DA Form 2946-R, *40-MM Grenade Launcher Scorecard.*

*The proponent for this publication is the US Army Training and Doctrine Command. The preparing agency is the US Army Infantry School. You may send comments and recommendations by any means, US mail, e-mail, fax, or telephone, as long as you use or follow the format of DA Form 2028, *Recommended Changes to Publications and Blank Forms*. You may also phone for more information:

E-mail	lu.lusano@us.army.mil
Phone	COM (706) 545-8623 or DSN 835-8623
Fax	COM (706) 545-8600 or DSN 835-8600
US Mail	Cmdt, USAIS ATTN: ATSH-INB 6650 Wilkin Drive , Bldg 74, Rm 102 Fort Benning, GA 31905-5593

*Unless this publication states otherwise, masculine nouns and pronouns may refer to either men or women.

CHAPTER 1

INTRODUCTION

This chapter discusses the training strategy and combat conditions for the 40-mm grenade launcher, M203. (Appendix A discusses the M79 model.)

* 1-1. TRAINING STRATEGY

An effective overall training strategy produces well-trained grenadiers and trainers by integrating resources into an effective year-round training program. Beginning with IET and continuing both in other institutions (NCOES, OBC, and CCC) and in the unit, such a program trains and sustains the individual and collective skills needed to perform the wartime mission. Specific training strategies are implemented by institutional and unit training programs, and supporting training strategies are implemented through the use of other resources such as publications, ranges, ammunition, training aids, devices, simulators, and simulations. The year-round program includes periodic preliminary marksmanship training followed by zeroing and range qualification firing. Other key elements of the program are training for the trainers and refresher training for nonfiring skills. The example in Figure 1-1 on page 1-3 shows the flow of unit sustainment training.

*a. **Institutional Training**. Training strategy begins with combat arms initial-entry training (IET), which trains soldiers in the standards of M203 gunnery tasks. Soldiers graduate with basic and advanced M203 skills that include maintaining the M203 and using it to hit a variety of targets. Other institutional training programs, such as NCOES, OBC, and CCC, reinforce these skills. Related soldier skills are integrated into tactical training (STP 21-1-SMCT).

b. **Unit Training**. Training continues in units, where, in addition to sustaining proficiency in skills gained in institutional training, leaders and soldiers develop and sustain new skills such as suppressive and supporting fire. These skills are integrated into collective training exercises to develop combat readiness. Preliminary marksmanship training is conducted before firings and as other opportunities arise. (Appendix B discusses an M203 unit training program.) To be effective, a unit training program focuses on three battlefield variables:

(1) ***Target***. Is the target moving or stationary, single or multiple?

(2) ***Grenadier***. Is the grenadier moving or stationary? Is he kneeling, prone, or standing?

(3) ***Conditions***. Is visibility full or limited? Must soldiers wear protective masks or not? Is it day or night?

c. **Initial and Sustainment Training**. A task that is taught correctly and learned well is retained longer, so initial training is critical. In addition to being more easily sustained, well-trained skills are also easier to regain if not used for some time. Retraining may be needed, however, if too much time elapses, if training doctrine changes, or if personnel turnover is high.

(1) ***Collective Training***. Collective training exercises progress from drills (squad, section, and platoon) to STXs, and then to live-fire tactical exercises (LFXs). Drill books

and MTPs provide tasks and guidance needed to plan and conduct the exercises. After each, leaders and trainers conduct an AAR to evaluate both individual and unit proficiency. The results provide readiness indicators and requirements for future training. LFXs provide leaders with an overview of unit proficiency and training effectiveness. They can be conducted on any range approved for M203 firing.

(2) ***Leader Training***. The most critical part of the Army's overall gunnery training strategy is to train the trainers and leaders first. Leader courses, however, include only limited M203 training, so officers and NCOs should use available publications to develop their proficiency with the M203. Publications help leaders plan, conduct, and evaluate their gunnery training programs. Proponent schools provide training support materials (field manuals, training aids, devices, simulators, and audiovisual programs), which provide the doctrinal foundations for training the force.

(3) ***Advanced Training***. Once the soldier knows the weapon and has demonstrated skill in zeroing, training strategy provides for additional live-fire training and target-acquisition exercises, which are conducted at various ranges. To develop proficiency, soldiers must master different types of targets and scenarios of increasing difficulty.

(4) ***Proficiency Assessment***. This is conducted on the zeroing and record live-fire exercise range when soldiers complete IET.

1-2. COMBAT CONDITIONS

The trainer must realize that qualification is not an end but a step toward reaching combat readiness. To reach combat readiness, the grenadier should consider his position, the capabilities of his weapon, and the following combat conditions:

a. Enemy personnel are seldom visible except when assaulting.

b. Most combat fire must be directed at an area where the enemy has been detected or is suspected but cannot be seen. Area targets consist of objects or outlines of men irregularly spaced along covered and concealed areas (ground folds, hedges, borders of woods).

c. Most combat targets can be detected by smoke, flash, dust, noise, or movement and are visible only for a moment.

d. Some combat targets can be engaged by using reference points, predetermined fire, or range card data.

e. The nature of the target and irregularities of terrain and vegetation may require a grenadier to use a variety of positions to place effective fire on the target. In a defensive situation, the grenadier usually fires from a supported fighting position.

f. Most combat targets have a low-contrast outline and are obscured. Therefore, choosing an aiming point in elevation is difficult.

g. Time-stressed fire in combat can be divided into three types:

- A single, fleeing target that must be engaged quickly.
- Area targets engaged within the time they remain available.
- A surprise target that must be engaged at once with instinctive, accurate fire.

*CHAPTER 2

DESCRIPTION AND MAINTENANCE

Proper weapon maintenance is a vital part of all gunnery training programs. Good maintenance contributes to weapon effectiveness as well as to unit readiness. This chapter provides a technical description of the M203 grenade launcher, its components, and its ammunition. It also discusses proper procedures for clearing, disassembling, cleaning, lubricating, inspecting, and caring for the weapon.

2-1. DESCRIPTION

The M203-series grenade launchers are lightweight, single-shot, breech-loaded, pump action (sliding barrel), shoulder-fired weapons (Figure 2-1).

a. The original model of the M203 was designed for use with M16-series rifles.

b. The M203A1 model was designed for use with M4 series carbines.

c. The M203A2 model has a quick-release bracket for use with M4 and M4A1 carbines that have the M4 rail system and with M16A4 rifles that have the M5 adapter rail system.

Model	Who Mounts/Dismounts	Weapon Mounted On
M203	Unit armorer	M16-series rifles
M203A1/A2	DS maintenance co	M4-series carbine rifles

NOTE: Individual soldiers should neither mount nor dismount the grenade launcher.

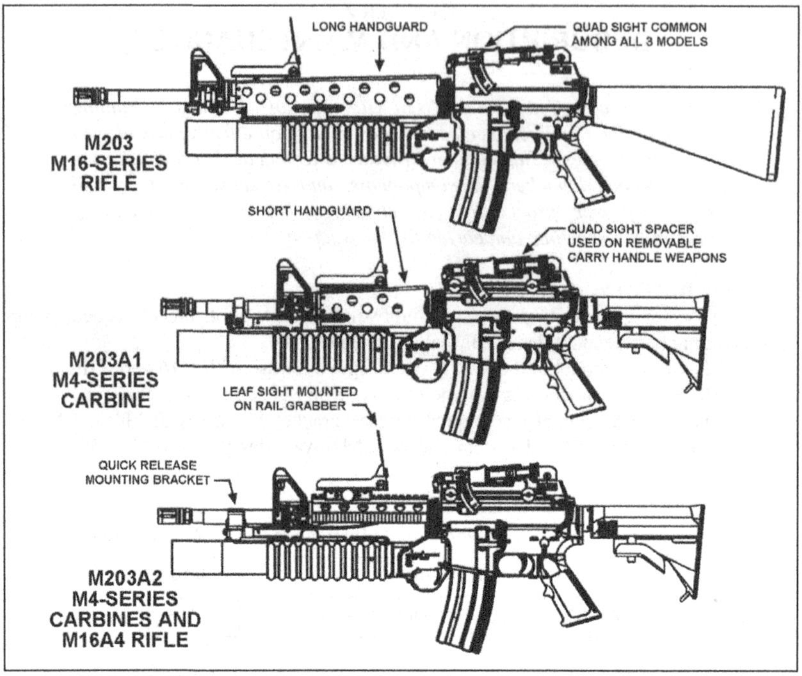

Figure 2-1. M203-series grenade launchers.

2-2. TECHNICAL DATA

Technical data for the M203-series grenade launchers follow:

a. **Weapon.**

Length	
Rifle and grenade launcher (overall)	99.0 cm (39 inches)
Barrel only	30.5 cm (12 inches)
Rifling	25.4 cm (10 inches)
Weight	
Launcher, unloaded	1.4 kg (3.0 pounds)
Launcher, loaded	1.6 kg (3.5 pounds)
Rifle and grenade launcher, both fully loaded	5.0 kg (11.0 pounds)
Number of lands	6 right hand twist

b. **Ammunition.**

Caliber	40 mm
Weight	About 227 g (8 oz)

c. **Operational Characteristics.**

Action	Single shot

Sights (use together or separately)	
Front leaf sight	Use between 50 and 250 meters (beyond 250 meters, the M16's barrel hides the target)
Rear quadrant sight	Use with M16's front sight post for ranges 50 to 250 meters.
Integrated front sight post, rear sight aperture	Use for 50- to 400 meters.
Chamber pressure	82,737 kilopascals (12,000 psi)
Muzzle velocity	76 mps (250 fps)
Maximum range	About 400 m (1,320 ft)
Maximum effective range	
Fire-team sized area target	350 m (1,155 ft)
Vehicle or weapon point target	150 m (495 ft)
Minimum safe firing range (HE)	
Training unprotected	165 m (541 ft)
Protected firing position	Less than 165 m (541 ft)
Combat	31 m (102 ft)
Minimum arming range	14 to 38 m (46 to 125 ft)
Rate of fire	5 to 7 rounds per minute
Minimum combat load	36 HE rounds

WARNING

When firing close-in, such as in urban areas, trenches, and other restrictive terrain, consider the minimum arming range to ensure the round clears other friendly forces.

2-3. COMPONENTS

Figure 2-2 shows the M203's major components, and the following paragraphs describe their purposes. Figure 2-3 through Figure 2-6 show the sight assemblies, the trigger and trigger guard, and the safety:

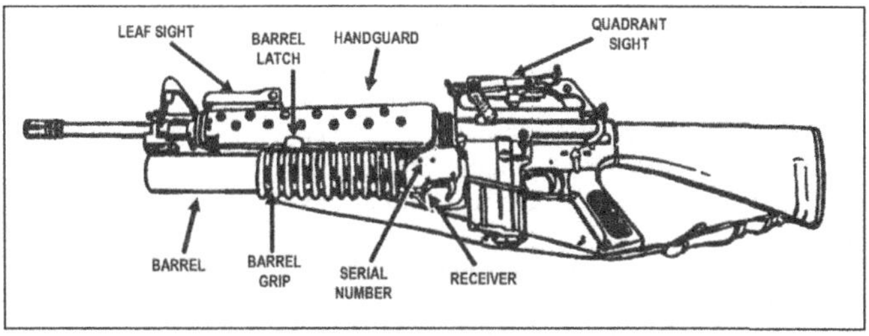

Figure 2-2. Components of the M203 grenade launcher.

a. **Handguard.** The handguard assembly houses the rifle barrel (Figure 2-2).

b. **Quadrant Sight Assembly**. The quadrant sight assembly, which attaches to the left side of the rifle's carrying handle, enables the grenadier to adjust for elevation and windage (Figure 2-2). This assembly consists of the sight, mounting screw, sight latch, rear sight aperture, sight aperture arm, front sight post, and sight post arm (Figure 2-3).

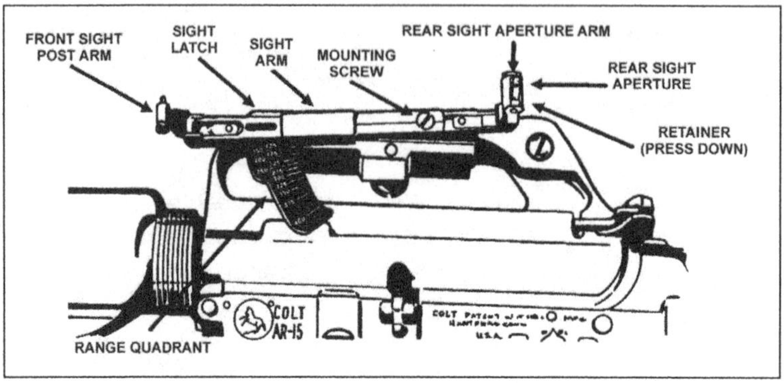

Figure 2-3. Quadrant sight assembly.

(1) ***Clamp, Bracket Assembly, and Mounting Screw***. The clamp and the bracket assembly hold the quadrant sight on the rifle's carrying handle. The mounting screw inserts through the right side of the clamp and into the bracket assembly.

(2) ***Sight Arm and Range Quadrant***. The sight arm mounts both the sight aperture arm (which holds the rear sight aperture) and the sight post arm (which holds the front sight post). This procedure allows the sight to pivot on the range quadrant to the desired range setting. The range quadrant is graduated in 25-meter increments from 50 to 400 m. Applying rearward pressure on the sight latch releases the quadrant sight arm so it can move along the range quadrant. Centering the number in the rear sight aperture selects the desired range. Releasing the sight latch locks the sight in position.

(3) ***Front Sight Post***. The front sight post mounts on the sight post arm by means of a pivot bracket. To prevent damage to the sights, keep the bracket closed when the sights are not in use. Use the sight post as follows to make minor adjustments in elevation when zeroing the launcher:

(a) To decrease elevation, turn the elevation adjustment screw on the sight post clockwise; to increase elevation, turn it counterclockwise.

(b) To move the impact of the projectile 5 meters at a range of 200 m, turn the elevation adjustment screw one full turn--360 degrees. To move the impact of the projectile 2.5 meters at a range of 200 m, turn the elevation adjustment screw one half turn--180 degrees.

(4) ***Rear Sight Aperture***. The rear sight aperture is on the sight aperture arm, which is attached to the rear portion of the quadrant sight arm. Use the rear sight aperture as follows to make minor adjustments in deflection (windage) when zeroing the launcher:

(a) To move the impact to the left, press the rear sight aperture retainer down and move the rear sight aperture away from the barrel; to move to the right, move it toward the barrel.

(b) To move the impact of the projectile 1.5 meter at a range of 200 m, move the rear sight aperture one notch.

c. **Receiver Assembly and Serial Number**. The receiver assembly houses the firing mechanism and ejection system and supports the barrel assembly. On the left side of the receiver assembly is the launcher's serial number (Figure 2-2).

d. **Barrel Assembly**. The barrel assembly holds the cartridges ready for firing and directs the projectile (Figure 2-2).

e. **Barrel Latch**. On the left side of the barrel is a latch that locks the barrel and receiver together (Figure 2-2). To open the barrel, depress the barrel latch and slide the barrel forward.

f. **Leaf Sight Assembly**. The leaf sight assembly is attached to the top of the handguard (Figure 2-4). The leaf sight assembly consists of the sight, its base and mount, an elevation adjustment screw, and a windage adjustment screw. Elevation and windage scales are marked on the mount. The folding, adjustable, open ladder design of the sight permits rapid firing without sight manipulation. The front sight post of the M16-series rifle serves as the front aiming post for the M203 leaf sight.

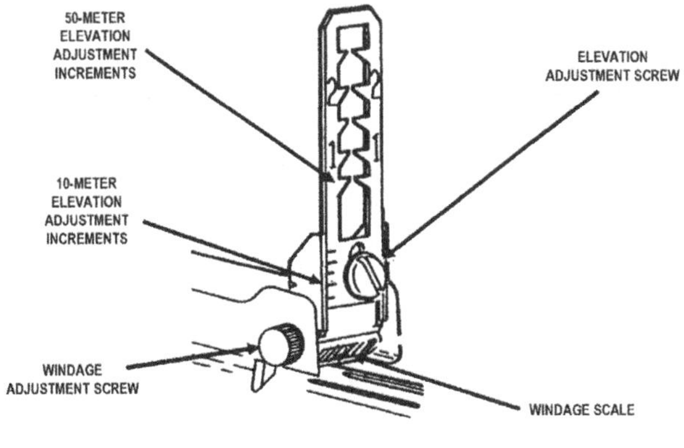

Figure 2-4. Leaf sight assembly.

(1) ***Sight Base***. Two mounting screws permanently attach the sight base to the rifle handguard. When the sight is down, the base protects it from damage.

(2) ***Sight Mount and Sight***. The grenadier uses the sight mount, which is attached to the sight base, to raise or lower the sight. Though the range is not marked on the sight in m, the sight is graduated in 50-meter increments from 50 to 250 m, which are marked with a "1" at 100 meters and a "2" at 200 m.

(3) ***Elevation Adjustment Screw and Elevation Scale***. The screw attaches the sight to its mount. When the screw is loosened, the sight can be moved up or down to make minor adjustments in elevation during the zeroing procedure. The rim of a 40-mm cartridge case is useful for turning the screw. Raising the sight increases the range;

lowering the sight decreases the range. The elevation scale consists of five lines spaced equally on the sight. The index line is to the left of the sight. Moving the sight one increment moves the impact of the projectile 10 meters in elevation at a range of 200 meters.

(4) ***Windage Screw and Windage Scale***. The firer uses the knob on the left end of the windage screw to make minor deflection adjustments during the zeroing. The scale has a zero line in the center. Two vertical lines are located on either side of and equidistant from the zero line. At a range of 200 meters, turning the knob on the windage scale one increment to the *left* moves the impact of the projectile 1.5 meter in the opposite direction, that is, to the *right*.

DANGER

NEVER **USE HIGH-EXPLOSIVE AMMO TO ZERO AT 50 METERS OR LESS--AT THAT RANGE, FRAGMENTATION COULD KILL YOU.**

TO REMIND YOU OF THIS, THE 50-METER MARK ON THE LEAF SIGHT BLADE IS PAINTED RED.

f. **Trigger Guard**. The trigger guard protects the trigger (A, Figure 2-5). Depressing the rear portion of the trigger guard rotates it down and away from the magazine well of the rifle. This gives the firer room to wear gloves or mittens while firing, if needed (B, Figure 2-5).

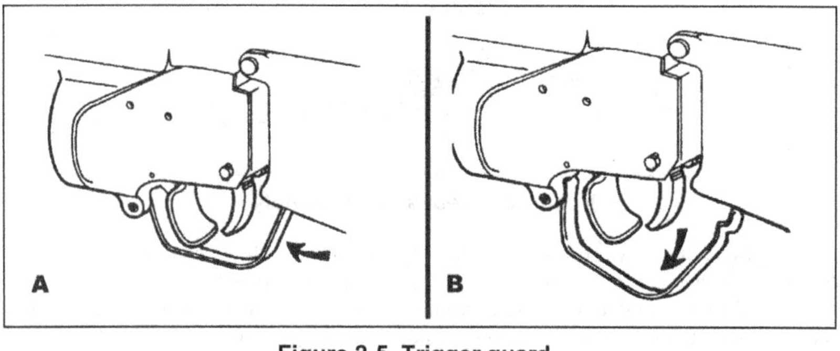

Figure 2-5. Trigger guard.

g. **Safety**. The safety is inside the trigger guard, just in front of the trigger. For the launcher to fire, the safety must be forward. When the safety is rearward, the launcher is on "S." The grenadier adjusts the safety manually (Figure 2-6).

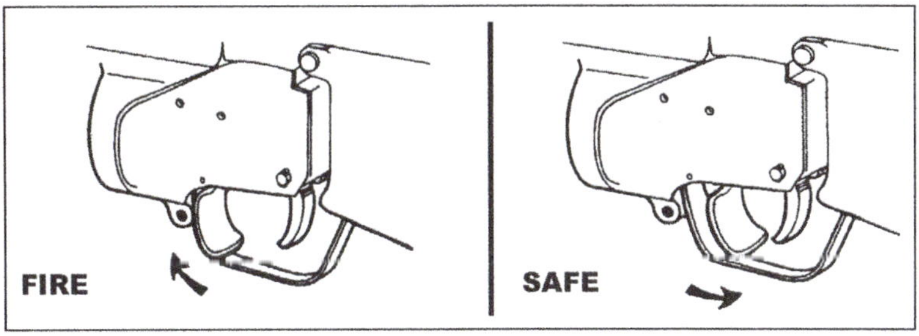

Figure 2-6. Safety.

2-4. AMMUNITION

The M203 grenade launcher uses several fixed-, low-velocity, 40-mm rounds. These include high-explosive, illuminating, signaling, CS, training, and multipurpose ammunition. This paragraph covers the most commonly used rounds.

> **WARNING**
>
> **If fired into snow or mud, 40-mm rounds might hit too softly to detonate. An undetonated round might then explode when stepped on or driven over.**
>
> **When training in snow or mud, avoid this hazard by firing only TP rounds.**

a. **Types, Characteristics, and Capabilities**. Figure 2-7 shows and Table 2-1 compares the M203 rounds (TM 43-0001-28 provides more details.)

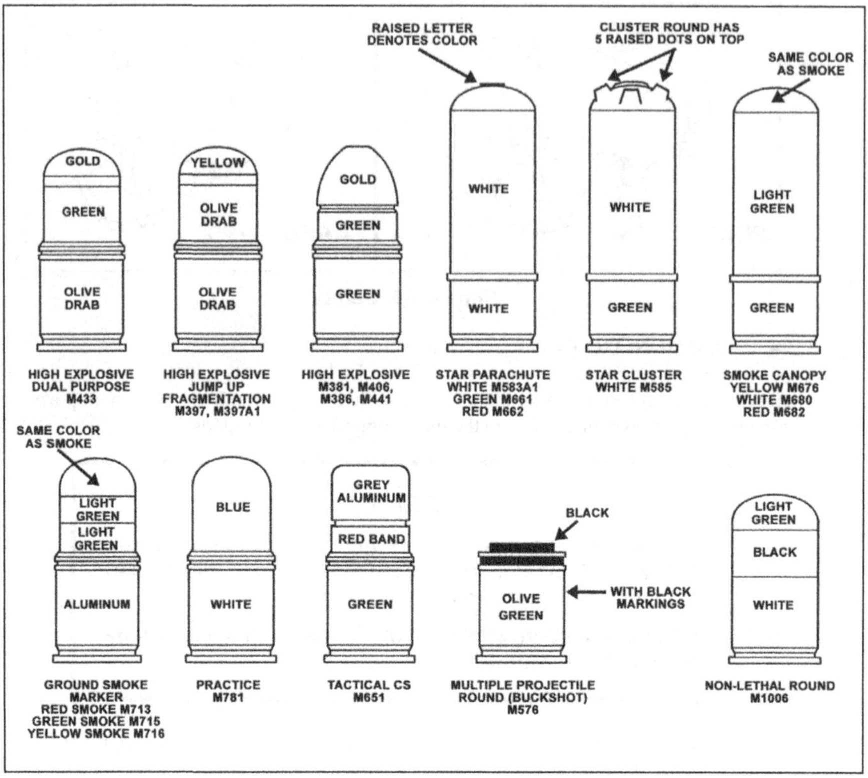

Figure 2-7. Cartridges for the M203 grenade launcher.

MODEL	DESCRIPTION	APPLICATION/EFFECTS
High Explosive, Dual-Purpose Round M433: DODAC 1310-B546 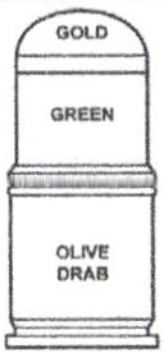	Ogive: Gold-colored steel cup Body: Green Skirt: Olive drab aluminum Markings: White Length: 10.29 cm (4.05 in) Weight: 0.23 kg (0.51 lb)	This round has a small slug (1/4 inch diameter) that penetrates at least 5 cm (2 in) of steel armor when fired straight (flat, that is, at 0 degree obliquity) within 150 m. When fired at a point target, it arms between 14 and 27 m (46 to 89 ft). It can cause-- • Casualties within 165 m. • Incapacitation within 15 m. • Death within 5 m.
WARNING **In training, seek cover before firing an HE round at a target within 165 meters.** **Also in training, never engage targets within a 165 meter radius, that is, within 165 meters in any direction from unprotected friendly troops.**		
High-Explosive Jump-Up Fragmentation M397 and M397A1 DODAC 1310-B569 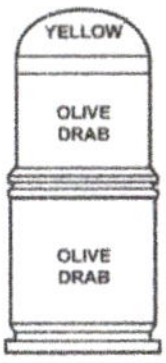	Ogive: Yellow steel ball Skirt: Olive drab aluminum Filler: Explosive Markings: Yellow Length: 4.05 in. Weight: 0.51 lb	These rounds are designed to inflict personnel casualties using an air burst effect. It arms between 14 and 27 meters (46 and 89 feet). It detonates and fragments about 5 feet above the ground impact point.
High-Explosive M381: DODAC 1310-B568 M441: DODAC 1310-B575 M386: DODAC 1310-B574 M406: DODAC 1310-B568 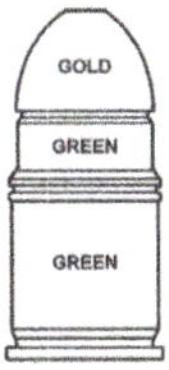	Ogive: Gold Body: Green Skirt: Olive drab aluminum with attached steel projectile Markings: Gold Length: 3.89 in. Weight: M381: 5.03 lb M441: 5.03 lb M386: 0.50 lb M406: 0.50 lb	These rounds produce a ground burst that causes casualties within a 165 meter radius. The kill radius is 5 meters. The M381 and M441 rounds arm within 2.4 to 3 meters (8 to 10 feet). The M386 and M406 rounds arm between 14 and 27 meters (46 to 89 feet).

Table 2-1. Comparison of M203 rounds.

MODEL	DESCRIPTION	APPLICATION/EFFECTS
Star Parachute Rounds M583A1 (WHITE) DODAC 1310-B531 90,000 CANDLEPOWER M661 (GREEN) DODAC 1310-B504 8,000 CANDLEPOWER M662 (RED) DODAC 1310-B505 20,000 CANDLEPOWER 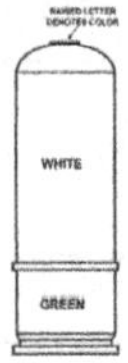	Cap: White, with raised letters indicating parachute color Body: White impact or bar alloy aluminum Skirt: Green Markings: Black Length: 13.39 cm (5.27 in) Weight: 0.22 kg (0.49 lb)	These rounds are used for illumination and signals, and are lighter and more accurate than comparable handheld signal rounds. The parachute attached to the round deploys upon ejection to lower the candle at 7 feet per second. The candle burns for about 40 seconds.
White Star Cluster M585 DODAC 1310-B536 	Cap: Plastic, with five raised dots for night identification. Body: White impact or bar aluminum alloy Skirt: Green Markings: Black Length: 13.38 cm (5.27 in) Weight: 0.19 kg (0.41 lb)	This round is used for illumination and signals. It is lighter and more accurate than comparable handheld signal rounds. The individual stars burn for about 7 seconds during free fall.
Ground Marker (Smoke) Rounds M713, (RED) DODAC 1310-B506 M715 (GREEN) DODAC 1310-B508 M716 (YELLOW) DODAC 1310-B509 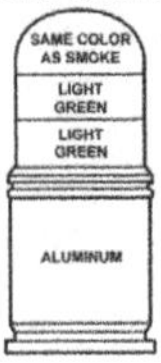	Ogive: Same as smoke color Body: Light green impact aluminum Skirt: Aluminum Markings: Black Length: 3.91 in. Weight: 0.49 lb	This round is used to mark soldiers' locations for others to see from the ground or the air. These rounds arm between 15 and 45 meters. If a fuze fails to function on impact, the output mixture in the front end of the delay casing backs up the impact feature.

Table 2-1. Comparison of M203 rounds (continued).

MODEL	DESCRIPTION	APPLICATION/EFFECTS
Smoke Canopy Rounds M676 (Yellow) DODAC 1310-B475 M680 (White) DODAC 1310-B477 M682 (RED) DODAC 1310-B479 	Ogive: Same as color of smoke Body: Light green impact aluminum Skirt: Aluminum Markings: Black Length: 5.212 in. Weight: 0.40 lb	This round is used to mark soldiers' locations for others to see from the ground or the air. These rounds arm between 15 and 45 meters. Burst height is 91 meters.
Practice Round M781 DODAC 1310-B519 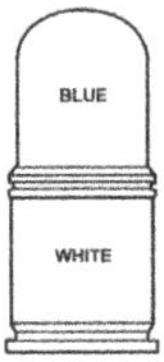	Windshield: Blue plastic Body: Zinc or aluminum Skirt: White Markings: White Length: 10.29 cm (4.05 in) Weight: 0.22 kg (0.48 lb)	Round produces a yellow or orange signature on impact.
Tactical CS M651 DODAC 1310-B567 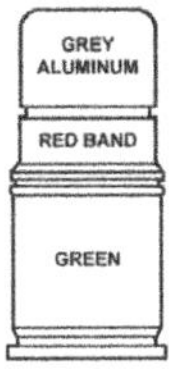	Ogive: Grey aluminum Band: Red Skirt: Green Markings: Black markings Length: 11.43 cm (4.50 in) Weight: 0.22 kg (0.48 lb)	This multipurpose round works best in riot control and urban operations. It arms between 10 to 30 meters (33 to 99 ft) and produces a white cloud of CS gas on impact.
Multiple Projectile (Buckshot) M576: DODAC 1310-B534 	Ogive: NA Body: Olive drab Skirt: NA Markings: Black Length: 6.72 cm (2.65 in) Weight: 0.12 kg (0.25 lb)	This multipurpose round is best in thick, vegetated areas or for use in room clearing. It contains at least 20 pellets, which cast a cone of fire 30 meters wide and 30 meters high, and which travel at 269 meters per second. Buckshot rounds should be aimed at the foot of the target. This round has no mechanical type fuze.

Table 2-1. Comparison of M203 rounds (continued).

MODEL	DESCRIPTION	APPLICATION/EFFECTS
Nonlethal Round M1006: DODAC 1310-BA06 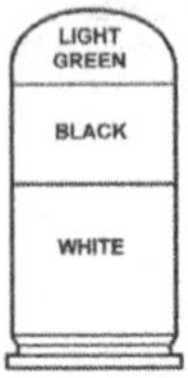	Nose: Green foam rubber Body: Black, high-density plastic Skirt: NA Other: Projectile assembly Cartridge case assembly Length: 3.95 in. Weight: 0.15 lb	This round is used to incapacitate personnel without penetrating the body. It is used to control riots, police, and keep the peace.

Table 2-1. Comparison of M203 rounds (continued).

b. **Storage**. Ammunition should be stored under cover. If this is not possible, it should be stored at least 15 centimeters (6 inches) aboveground, and covered with a double layer of tarpaulins. Tarpaulins should protect the ammunition, but allow ventilation. Trenches should be dug to prevent water from flowing under the ammunition.

c. **Care, Handling, and Preservation**. Ammunition containers should remain tightly closed until the ammunition is needed. Ammunition removed from the airtight container is likely to corrode, particularly in damp climates. Soldiers must also take the following precautions:

(1) ***Protect Ammunition from Mud, Dirt, and Water***. If the ammunition gets wet or dirty, wipe it off before use. Also, wipe off lightly corroded cartridges as soon as you find the corrosion. Never fire heavily corroded or dented projectiles or those with loose parts or particles.

> **WARNING**
> **Never fire heavily corroded or dented projectiles, or those with loose parts or particles.**

(2) ***Avoid Exposing Ammunition to the Direct Rays of the Sun.*** Hot powder can cause excessive pressure when the round is fired.

(3) ***Do Not Lubricate Ammunition.*** This can cause dust and other abrasives to collect on it and damage the operating parts of the launcher.

d. **Packaging**. Ammunition packaging varies according to the type of ammunition:

(1) ***HE, HEDP, and TP***. Each wire-bound box of HE and HEDP ammunition contains 1 fiber box with 12 bandoleers of 6 rounds each, for a total of 72 rounds. Each wire-bound box of M781 practice rounds contains 4 fiber boxes of 25 rounds each, for a total of 100 rounds.

(2) ***Smoke and Cluster Ammunition***. Each wire-bound box of smoke and cluster ammunition contains 2 cans with 22 rounds each, for a total of 44 rounds.

(3) ***CS Ammunition***. Each box of CS ammunition contains 2 cans with 4 bandoleers of 6 rounds each, for a total of 48 rounds.

(4) ***Buckshot***. Each box of buckshot ammunition contains 12 bandoleers of 6 rounds each, for a total of 72 rounds.

(5) ***Nonlethal Ammunition.*** Each wire-bound wooden box contains 22 rounds per M2A1 metal box, for a total of 44 rounds.

3-5. CLEARING PROCEDURES

The soldier must clear the weapon before performing maintenance on it. FM 3-22.9 (23-9) provides instructions for clearing an M16-series rifle. To clear the grenade launcher--

a. Push in the release button and pull the barrel forward.
b. Watch to see if a round extracts.
c. Place the safety on "S."
d. Inspect the breech to ensure a round is not present.
e. Pull the barrel to the rear until it clicks. This cocks the weapon.

2--6. GENERAL DISASSEMBLY

As he disassembles the weapon, the soldier places each part on a clean, flat surface such as a table, shelter half, or disassembly mat. This aids in reassembly and simplifies the task of keeping up with the parts. The soldier will later reassemble the grenade launcher in the reverse order that he disassembled it. (Only ordnance personnel may disassemble the grenade launcher beyond the steps described here.) To disassemble the weapon--

a. Loosen the mounting screw and remove the quadrant sight assembly from the carrying handle of the M16-series rifle (Figure 2-8).

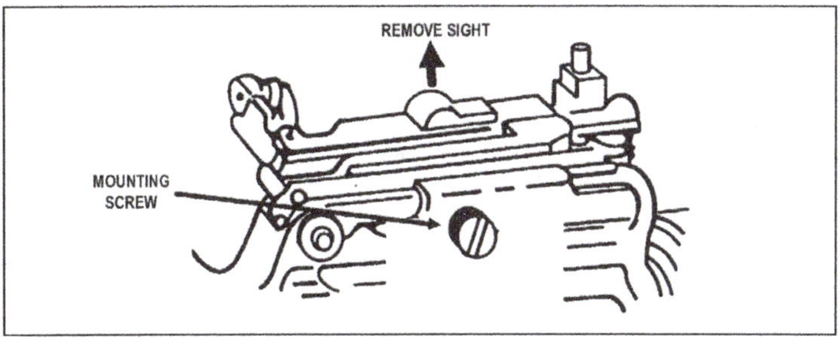

Figure 2-8. Removal of the quadrant sight assembly.

b. Remove the barrel assembly and handguard assembly, in either order:

(1) ***Barrel Assembly First***. Push the barrel latch and move the barrel forward until it hits the barrel stop. On the left side of the handguard, insert a cleaning rod into the fourth hole back from the muzzle, depress the barrel stop, and slide the barrel forward and off (Figure 2-9).

Figure 2- 9. Removal of the barrel assembly before the handguard assembly.

(2) ***Handguard Assembly First***. Pull back on the M16's slip ring and remove the handguard by pulling it up and back. Push the barrel latch and move the barrel forward until it hits the barrel stop. Use a cleaning rod to depress the barrel stop and slide the barrel forward and off (Figure 2-10).

> **CAUTION**
> Never use a screwdriver or any other tool when removing the hand guard assembly. Doing so could damage the hand guard assembly or slip ring.

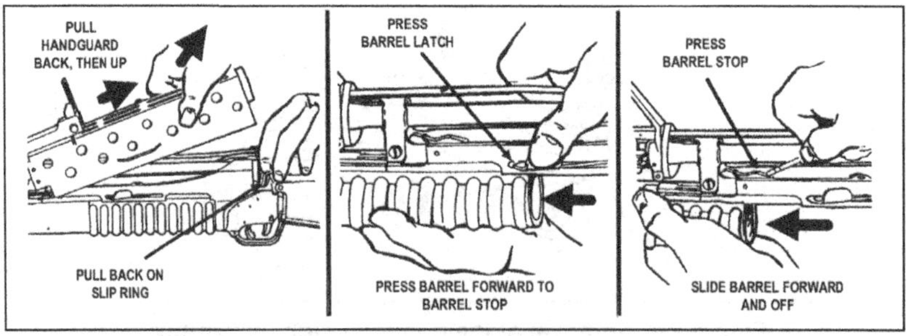

Figure 2-10. Removal of the handguard assembly before the barrel assembly.

2-7. CLEANING AND LUBRICATING PROCEDURES

After firing the grenade launcher, or if it has been idle for a long time, the soldier must clean and lubricate it as follows:

a. **Bore**. Attach a clean, dry rag to the thong, and then thoroughly moisten the rag with CLP. Pull the rag through the bore several times. Attach the bore brush to the thong, pull it through the bore several times, and follow this with more rags moistened with CLP (Figure 2-11). Pull dry rags through the bore, and inspect each rag as it is removed. The bore is clean when a dry rag comes out clean. Finally, pull a rag lightly moistened with CLP through the bore to leave a light coat of lubricant inside the barrel.

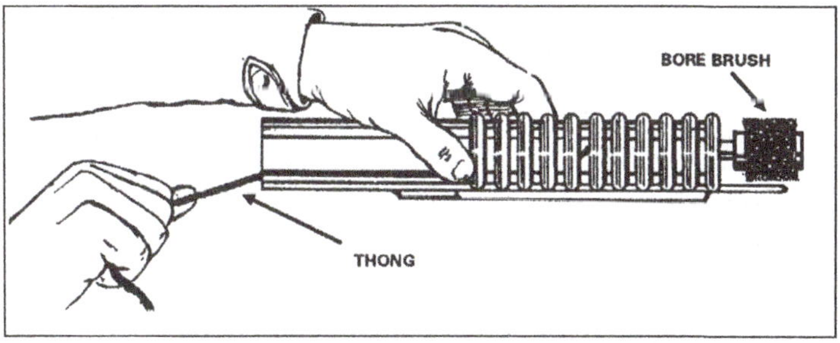

Figure 2-11. Cleaning of the barrel with thong and bore brush.

b. **Breech Insert**. Clean the face of the breech insert with a patch and CLP. Remove the CLP with dry rags; then lubricate the breech with a new, light coat of CLP.

c. **Other Parts**. Use a brush and dry rags to clean all the other parts and surfaces. After cleaning, apply a light coat of CLP to the outside of the launcher.

d. **Safety Mechanism**. Clean the safety mechanism properly with CLP; then lubricate it with CLP.

e. **Special Lubrication Requirements**. Lubricate the grenade launcher only with CLP and IAW the following environmental guidelines:

(1) ***Extreme Heat***. Lubricate with CLP, grade 2.

(2) ***Damp or Salty Air***. Clean the weapon and apply CLP, grade 2, frequently.

(3) ***Sandy or Dusty Air***. Clean the weapon and apply CLP, grade 2, frequently. Remove excess CLP with a rag after each application.

(4) ***Temperatures Below Freezing***. When the weapon is brought in from a cold area to a warm area, keep it wrapped in a parka or blanket, and allow it to reach room temperature gradually. If condensation forms on the weapon, dry and lubricate it at room temperature with CLP, grade 2, before returning it to cold weather. Otherwise, ice will form inside the mechanism.

NOTE: Although CLP provides sufficient lubrication down to -35°F (-37°C), it will only flow from its 1/2-ounce bottle, at 0°F (-17°C) and above.

2-8. INSPECTION

Inspection begins with the weapon already disassembled into its major groups or assemblies. Parts with shiny surfaces are serviceable. The following parts of the weapon and related equipment are inspected IAW TM 9-1010-221-10:

a. **All Parts**. Check for wear and damage, including burrs, scratches, and nicks.

b. **Handguard**. Check for cracks, dents, or distortion that prevents its firm attachment to the rifle.

c. **Leaf Sight Assembly**. Check for bent or damaged parts, rust or corrosion, and legible markings.

d. **Barrel**. Check for cracks or dents.

e. **Cartridge and Retainers**. Check for breakage, bends, chips, or missing parts.

NOTE: Take any unserviceable part to the armorer, who will decide if it is serviceable and replace parts as needed.

2-9. GENERAL ASSEMBLY

The soldier assembles the grenade launcher in the reverse order of disassembly.

a. Install the barrel by pressing the barrel stop and sliding the barrel into the receiver (Figure 2-12).

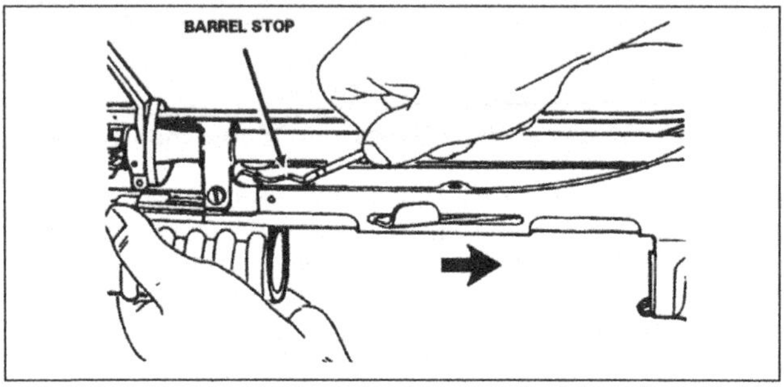

Figure 2-12. Installation of the barrel.

b. Lock the barrel by moving it rearward until it closes with a click (Figure 2-13).

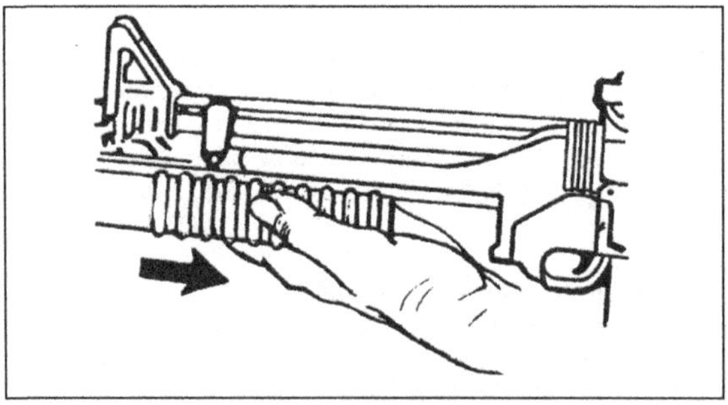

Figure 2-13. Locking of the barrel.

c. Install the handguard and secure it with the slip ring (Figure 2-14).

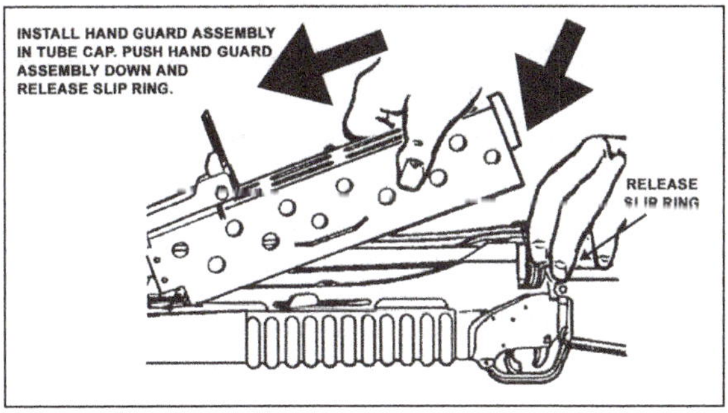

Figure 2-14. Installation and securing of the handguard.

d. Install the quadrant sight assembly (Figure 2-15).

Figure 2-15. Installation of the quadrant sight assembly.

e. Perform a function check to ensure that the grenade launcher has been assembled correctly. Notify the unit armorer at once if the launcher fails to function. Conduct the function check in this order:

(1) Check the proper operation of the sear. Cock the launcher, pull the trigger, and continue to hold the trigger. When you pull the trigger, you release the firing pin and should hear a metallic click The firing pin should release with a metallic click. Hold the trigger to the rear and cock the launcher again. Release the trigger, then pull. The firing pin should again release.

> **WARNING**
>
> **If the sear malfunctions, the launcher could fire without your pulling the trigger.**

(2) Check the safety by pulling the trigger in both the "S" and "F" positions. You must cock the launcher before you can place the safety in the "S" position.

(3) Check the windage adjustment screw on the leaf sight assembly for proper operation.

(4) Move the elevation adjustment screw only if the weapon has been zeroed.

(5) Move the barrel forward and back to be sure the barrel stop and barrel latch function.

2-10. CARE AND HANDLING PROCEDURES

Certain steps must be taken before, during, and after firing to properly maintain the grenade launcher.

a. **Before firing**.

- Wipe the bore dry.
- Inspect the weapon as outlined in the operator's technical manual.
- Ensure the weapon is properly lubricated.

b. **During firing**.

- Periodically inspect the weapon to ensure that it is lubricated.
- When malfunctions or stoppages occur, see the procedures in Chapter 4.

c. **During CBRN Conditions**. If contamination is anticipated, the soldier should apply CLP to all outer metal surfaces of the weapon. Ammunition, however, should never be lubricated. The soldier should keep the weapon covered as much as possible. If the weapon is contaminated, he should first decontaminate it IAW FM 3-3 and FM 3-11.4, and then he should clean and lubricate it.

2-11. DECONTAMINATION

Leaders must try to reduce the penetration of contaminants and exposure to them. Contaminated material is disposed of IAW SOP.

a. **Chemical**. Use towelettes from the M258A1 kit to wipe off the weapon. If these are unavailable, wash the weapon with hot soapy water.

b. **Biological**. Use towelettes from the M258A1 kit to wipe off the weapon. If these are unavailable, wash with soap and water.

c. **Radiological or Nuclear.** Wipe off the weapon with warm soapy water. Otherwise, use towelettes or rags (FM 3-11.5).

*CHAPTER 3
OPERATION AND FUNCTION

This chapter discusses the operation and function of the M203 grenade launcher.

3-1. OPERATION

The grenadier's operations include loading, unloading, and firing the weapon. The weapon uses a high-low propulsion system to fire a 40-mm round. The firing pin strikes the primer, whose flash ignites the propellant in the brass powder-charge cup inside the high-pressure chamber. The burning propellant produces 12,000 psi chamber pressure, which ruptures the brass powder-charge cup at the vent holes and allows the gases to escape to the low-pressure chamber in the cartridge case. There the pressure drops to 3,000 psi and propels the grenade from the muzzle at a velocity of 250 fps. The grenade's 3,700-rpm right-hand spin stabilizes it during flight and applies enough rotational force to arm the fuze. The weapon is unloaded with the barrel open and fired from a closed bolt. It must be cocked before it can be placed on safe.

3-2. LOADING OF M203 WITH THE SAFETY ON ("S")

When the safety is on "S," the grenadier loads the M203 as follows (Figure 3-1):

a. Keep the safety on "S" until ready to fire.

b. Depress the barrel latch.

c. Slide the barrel assembly forward. This automatically extracts a round and cocks the weapon.

CAUTION

Before loading the M203, make sure that the bore and chamber are clean and dry.

d. Insert clean, dry, undented ammunition into the chamber.

e. Slide the barrel rearward until it locks with an audible click.

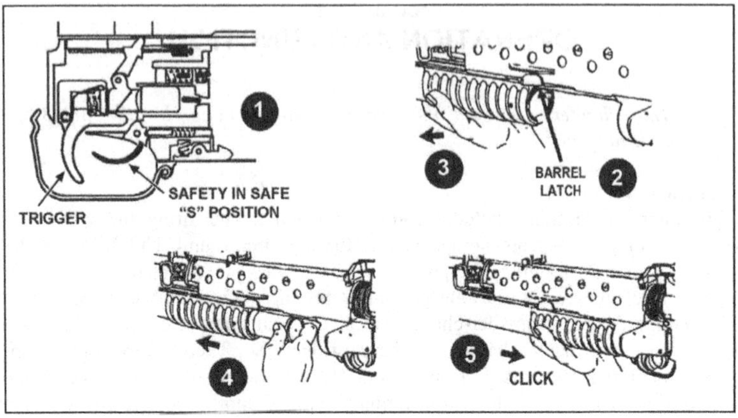

Figure 3-1. Loading the M203 with the safety on ("S").

3-3. LOADING OF M203 WITH THE SAFETY OFF ("F")

When the safety is off, the grenadier loads the M203 as follows (Figure 3-2):

a. Depress the barrel latch.

b. Slide the barrel assembly forward. This automatically extracts a round and cocks the weapon.

c. Place the weapon on safe ("S").

CAUTION

Before loading ensure bore and chamber is clean and dry.

d. Insert clean, dry, undented ammunition into the chamber.

e. Slide the barrel rearward until it locks with an audible click.

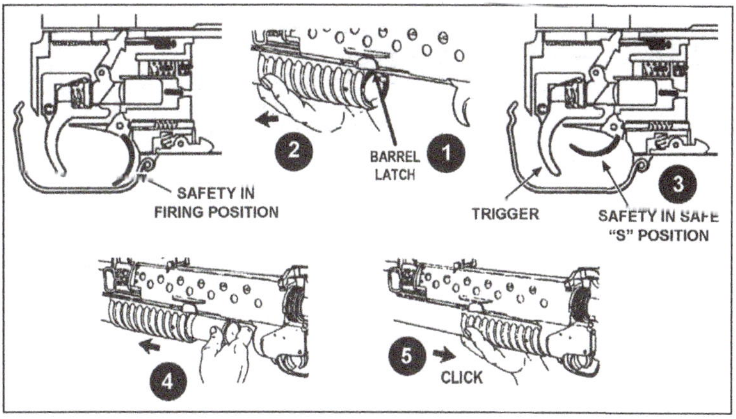

Figure 3-2. Loading the M203 with the safety off ("F").

> **WARNING**
>
> **Keep the muzzle pointed downrange and clear of all Soldiers. Use the correct ammunition:**
>
> - **Never fire pyrotechnic ammunition made for the AN-M8 pyrotechnic pistol in the M203.**
> - **Never load aircraft ammunition M384 (HE) or M385 practice.**
> - **Never use high-velocity 40-mm ammunition designated for other 40-mm weapons such as the MK 19. High-velocity rounds are longer than those used in the M203 and might cause this weapon to explode, resulting in serious injury.**

3-4. UNLOADING OF M203

To unload the grenade launcher, the grenadier must first depress the barrel latch and move the barrel forward. The cartridge case or round should automatically eject. If the case sticks, the grenadier taps it with a cleaning rod to remove it (Figure 3-3). He places the weapon on safe ("S"), then slides the barrel rearward, locking it to the breech.

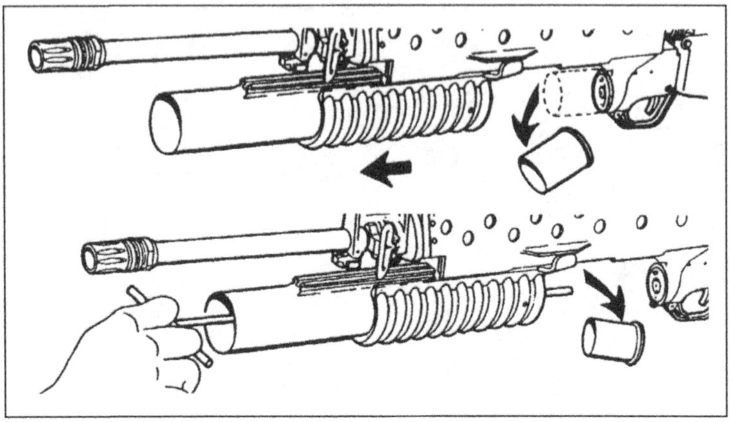

Figure 3-3. Unloading the M203 grenade launcher.

WARNING

To unload an unfired weapon, avoid detonation by either--

- **Catching the ejected round or**
- **Holding the weapon close to the ground to reduce the distance the round can fall.**

3-5. CYCLE OF FUNCTIONING

Knowing the M203's cycle of functioning from loading to firing helps grenadiers recognize and correct stoppages. Many of the actions described in this chapter occur at once, but here they are explained separately.

a. **Unlocking**. The cycle begins when the grenadier depresses the barrel latch to unlock the barrel assembly and slides the barrel assembly forward (Figure 3-4).

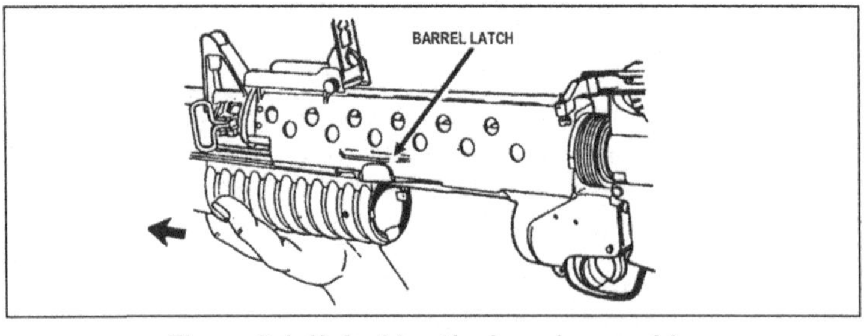

Figure 3-4. Unlocking the barrel assembly.

b. **Cocking**. The grenadier moves the barrel assembly forward, then backward, to cock the weapon. As the barrel assembly moves, it takes with it the barrel extension. Their movement causes the following to occur:

(1) The cocking lever is forced down as the barrel assembly and barrel extension, which are interlocked with the cocking lever, move forward.

(2) The movement of the cocking lever forces the spring-loaded firing pin to the rear.

(3) The spring-loaded follower also moves forward with the barrel extension.

(4) The barrel assembly continues forward, disengaging the barrel extension from the cocking lever. The cocking lever is then held down by the follower.

(5) When the grenadier begins to move the barrel assembly back to the rear, this forces the follower to the rear.

(6) The cocking lever again engages the barrel extension, which causes the firing pin to move slightly forward and engage the primary trigger sear. This cocks the weapon (Figure 3-5).

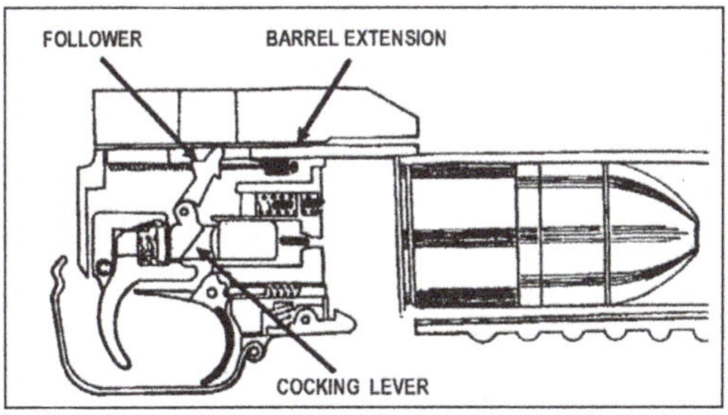

Figure 3-5 . Cocking the M203 grenade launcher.

c. **Extracting**. Extracting and cocking occur at the same time. As the grenadier opens the barrel assembly, a spring-loaded extractor keeps the live round or spent cartridge case seated against the receiver until the barrel clears the cartridge case (Figure 3-6).

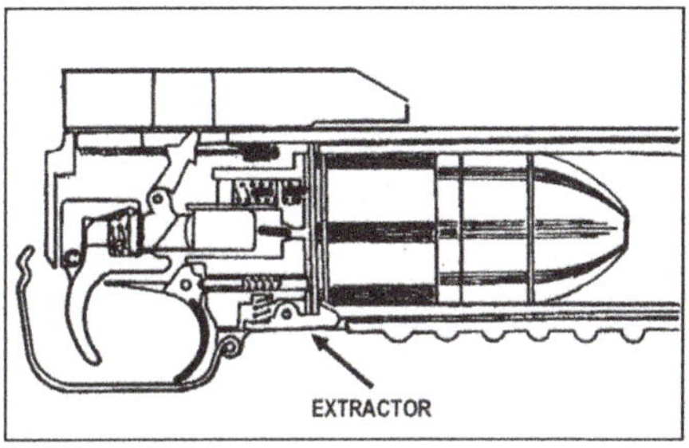

Figure 3-6 . Extracting the round or cartridge case.

d. **Ejecting**. The spring-loaded ejector pushes the live round or spent cartridge case from the barrel assembly (Figure 3-7).

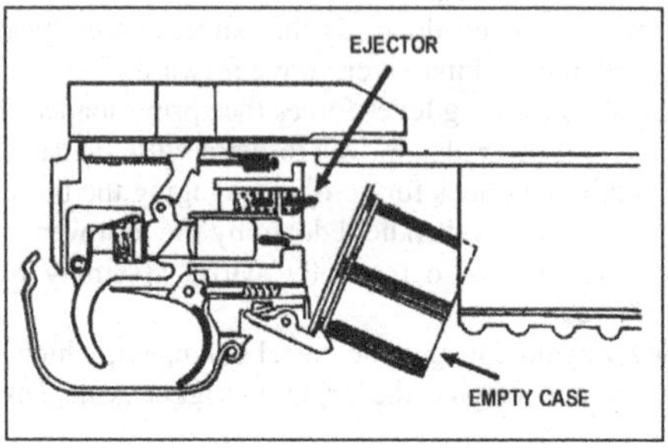

Figure 3-7 . Ejecting the round or cartridge case.

e. **Loading**. With the barrel assembly open, the grenadier inserts a round into the breech end of the barrel (Figure 3-8).

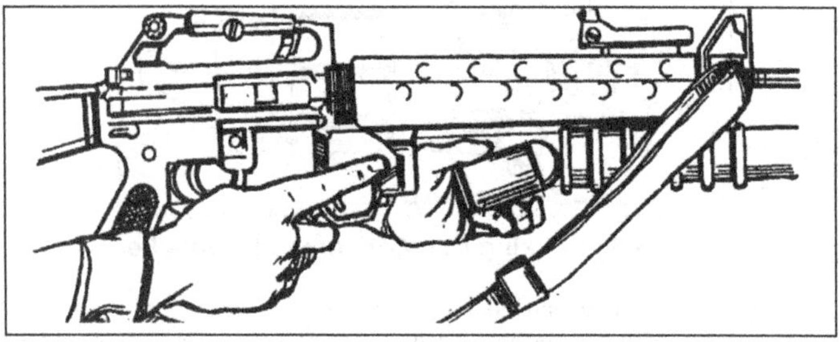

Figure 3-8 . Loading the M203 grenade launcher.

f. **Chambering**. As the grenadier closes the breech end of the barrel assembly, the extractor contacts the rim of the cartridge and seats (chambers) the round firmly (Figure 3-9).

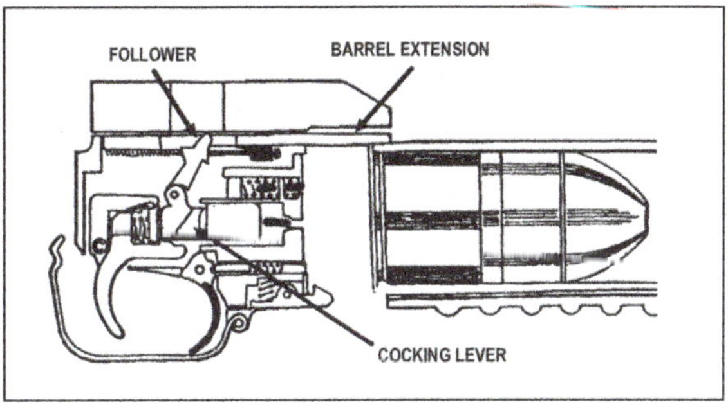

Figure 3-9 . Chambering a round.

g. **Locking**. As the barrel assembly closes, the barrel latch engages it. The cocking lever engages the barrel extension so that it cannot move forward along the receiver assembly.

h. **Firing**. When the grenadier pulls the trigger, the primary trigger sear disengages from the bottom sear surface of the firing pin. This releases the spring-driven firing pin, forcing it forward against the cartridge primer (Figure 3-10).

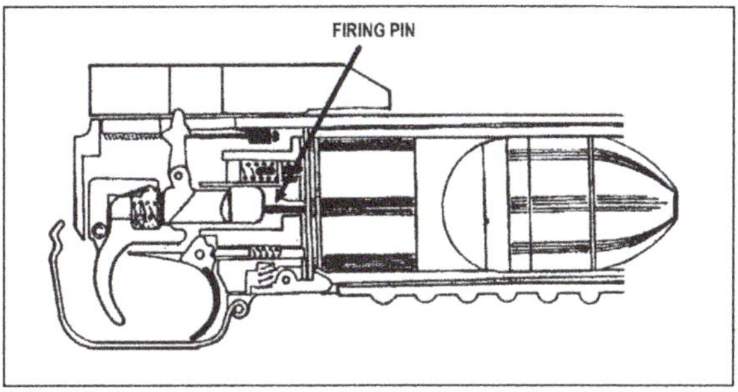

Figure 3-10. Firing the M203 grenade launcher.

This page intentionally left blank.

CHAPTER 4

PERFORMANCE PROBLEMS AND DESTRUCTION

This chapter identifies some of the problems that can cause the M203 grenade launcher to perform incorrectly. It also explains how to identify unserviceable parts and how to destroy the weapon when authorized to do so.

4-1. MALFUNCTIONS

A malfunction occurs when a mechanical failure prevents the weapon from firing properly. Neither defective ammunition nor improper operation of the weapon by the firer is a malfunction. The weapon should be cleaned, lubricated, and retried. If it still fails to function, it should be turned in to the unit armorer. Table 4-1 shows probable causes and corrective action for each type of malfunction.

<table>
<tr><th>Malfunction</th><th>Probable Cause</th><th>Corrective Action</th></tr>
<tr><td rowspan="3">Failure to cock</td><td>Broken sear</td><td rowspan="4">Notify unit maintenance</td></tr>
<tr><td>Improper assembly of cocking lever</td></tr>
<tr><td>Loose, broken, or missing cocking lever spring pin</td></tr>
<tr><td>Failure to lock</td><td>Excess plastic on breech end of barrel assembly</td></tr>
</table>

Table 4-1. Malfunctions.

4-2. STOPPAGES

A stoppage is an unintentional interruption in the cycle of operation or functioning that may be cleared by immediate action. A stoppage is classified by its relationship to the cycle of functioning. Table 4-2 on page 4-2 shows the types of stoppages.

Stoppage	Probable Cause	Corrective Action
Failure to fire	Safety on	Place in fire position
	Empty chamber	Load
	Faulty ammunition	Reload
	Water or excess lubricant in firing pin well	Hand cycle weapon several times, to include pulling the trigger
	Worn or broken firing pin	Notify unit maintenance
	Dirt or residue in firing pin recess	Clean
	Blurred sear or firing pin	Notify unit maintenance
	Dirty firing pin well opening	
	Weak or broken firing pin spring	
Failure to extract	Defective extractor on spring or spring pin	
	Ruptured cartridge case	Remove from barrel
Failure to eject	Worn, broken, or missing ejector spring or retainer	Notify unit maintenance
Failure to chamber	Faulty ammunition	Reload
	Dirty chamber	Clean bore and chamber
Safety fails to stay in position	Missing spring pin or broken or worn safety	Notify unit maintenance

Table 4-2. Stoppages.

*4-3. IMMEDIATE ACTION

Immediate action refers to anything a Soldier does to reduce a stoppage without taking time to look for the cause. Immediate action should be taken in the event of either a hangfire or misfire. Either can be caused by an ammunition defect or by a faulty firing mechanism. Any failure to fire must be considered a hangfire until that possibility is eliminated.

a. What is the difference between a hangfire, a misfire, and a stoppage?

(1) A *hangfire* is a delay in the functioning of the round's propelling charge explosive train at the time of firing. The length of this delay is unpredictable, but in most cases, it ranges between a split second and 30 seconds. Such a delay in the functioning of the round could result from the presence of excess oil or grease, grit, sand, frost, or ice.

(2) A *misfire* is a complete failure of the weapon to fire. A misfire in itself is not dangerous, but because it cannot be immediately distinguished from a hangfire, it must be considered to be a hangfire until proven otherwise.

(3) A *stoppage* may be caused by a hangfire, the following precautions must be observed until the round has been removed from the weapon and the cause of the failure determined:

(a) Keep the M203 pointed downrange or at the target and keep everyone clear of its muzzle. If the stoppage occurs during training, shout MISFIRE and clear the area of any Soldiers not needed for the operation.

(b) Wait 30 seconds from the time of the failure before opening the barrel assembly. Once the 30 seconds has elapsed, open the barrel assembly slowly to about one-half inch or until you hear a metallic click. As soon as you hear the click, place the weapon on safe. You are now ready to perform the unloading procedure.

(c) After removing the round from the receiver, determine whether the round or the firing mechanism is defective. Examine the primer to see if it is dented. If the primer is dented, separate the round from other ammunition until it can be disposed of properly. However, if the primer is not dented, the firing mechanism is at fault. Once the cause of the failure to fire has been corrected, the round may be reloaded and fired.

WARNING

If you are unloading a weapon that has not been fired, avoid detonation either by catching the ejected round or by holding the weapon close to the ground to reduce the distance the round can fall.

4-4. REMEDIAL ACTION

Remedial action is any action taken by the gunner to restore his weapon to operational condition. Take remedial action only if immediate action does not remedy the problem.

4-5. DESTRUCTION PROCEDURES

Destruction of any military weapon is authorized only as a last resort to prevent the enemy from capturing or using it. This paragraph discusses planning for destruction, priorities and methods of destruction, and degree of damage. In combat situations, the commander has the authority to destroy weapons, but he must report doing so through channels.

a. **Planning**. SOPs for all units should contain a plan for destroying equipment. Having such a plan ensures that the damage is effective enough to deny use of the equipment to the enemy. The plan must be flexible enough in its designation of time, equipment, and personnel to meet any situation.

b. **Priorities of Destruction**. When lack of time prevents them from completely destroying equipment, Soldiers must destroy the same essential parts on all like equipment. The order in which the parts should be destroyed (priority of destruction) follows:

(1) Bolt assembly (M16) and breech mechanism (M203).

(2) Barrels (both M16 and M203).

(3) Sights or sighting equipment (including nightsight).

(4) Optics mount.

c. **Methods of Destruction**. Equipment may be destroyed by any of several methods. The commander must use his imagination and resourcefulness to select the best method of destruction based on the facilities available. Time is usually critical. The methods of destruction are as follows:

(1) ***Mechanical.*** Use an axe, pick, sledgehammer, crowbar, or other heavy implement.

(2) ***Burning.*** Use gasoline, oil, incendiary grenades, other flammables, or a welding or cutting torch.

(3) ***Demolition.*** Use suitable explosives or ammunition or, as a last resort, hand grenades.

(4) ***Disposal.*** Bury essential parts, dump them in streams, or scatter them so widely that recovering them would be impossible.

d. **Degree of Damage**. The method of destruction used must damage equipment and essential spare parts to the extent that they cannot be restored to usable condition in the combat zone, either by repair or by cannibalization.

CHAPTER 5

MARKSMANSHIP TRAINING

**Marksmanship training is conducted in three phases. Every phase has the same three objectives: to teach each grenadier--*

- *To hit the target accurately with the first round,*
- *To adjust fire, and*
- *To do both quickly.*

This chapter discusses the first two phases: ***preliminary marksmanship training****, which develops nonfiring individual skill proficiency (Section I), and* ***basic gunnery*** *(Section II). In these two phases, the Soldier learns to apply the fundamentals of gunnery and to zero the M203 during qualification exercises in day, CBRN, and night conditions. (Chapter 6 discusses the third phase,* ***advanced gunnery****.)*

***WARNING**

Before allowing anyone to move between stations, ensure that--

- **All rifles and grenade launchers have been cleared.**
- **Bolts are to the rear.**
- **Barrel assemblies are in the open position.**

Anyone observing an unsafe act should call out "Cease fire" and notify range personnel immediately.

Section I. PRELIMINARY MARKSMANSHIP TRAINING

Grenadiers and leaders must master marksmanship fundamentals before firing individually or collectively. During preliminary marksmanship training, grenadiers learn and demonstrate the individual skills that prepare them to fire live ammunition. After learning the characteristics and mechanics of the weapon (Chapters 2, 3, and 4), they learn the four fundamentals of marksmanship, sight manipulation, and response to fire commands. Dry-fire exercises are excellent for training to proficiency. Good preliminary marksmanship instruction improves individual proficiency, which in turn improves the proficiency of collective fire.

*5-1. FOUR FUNDAMENTALS OF MARKSMANSHIP

The four fundamentals of M203 marksmanship are steady position, aiming, breathing, and trigger control. Only the first fundamental (steady position) varies. The other three remain the same regardless of the Soldier's position.

a. **Steady Position**. This varies according to the position and the type of sight used (quadrant or leaf). Note that for firing prone, a supported position is best.

(1) ***Prone Position, Quadrant Sight*** (top, Figure 5-1).

(a) Lie face down, grasp the M16 pistol magazine with your right hand, and place the butt of the rifle into the pocket of your right shoulder.

(b) Lower your right elbow to the ground so your shoulders are level. This places the weight of your body behind the weapon, which enables you to recover quickly each time you fire.

(c) Grasp the barrel grip with your left hand, supporting with sandbags. Straighten your upper body and spread your legs a comfortable distance apart. Try to point your toes outward and relax your ankles so your heels will rest on the ground. Relax the weight of your upper body forward onto your left arm.

(2) ***Prone Position, Leaf Sight*** (bottom, Figure 5-1). While firing with the leaf sight at ranges greater than 150 meters, place the butt stock of the weapon under your armpit and grip firmly to prevent the weapon from moving. Lean your head 45 degrees to the right and place the M16 front sight post on the desired range. Raise the butt stock and lower the muzzle to obtain the proper sight alignment and sight picture.

Figure 5-1. Prone supported position, quadrant sight (top) and leaf sight (bottom).

***WARNING**

Before allowing anyone to move between stations, ensure that the sling is clear of the weapon muzzle before firing.

(3) ***Kneeling Position*, *Quadrant Sight*** (left, Figure 5-2).

(a) Kneel on your right knee while facing the target with your right hand on the magazine and your left hand grasping the barrel grip.

(b) Place your left foot about .45 meter (18 inches) to your left front with your toes pointing in the general direction of the target.

(c) Keeping your right toe in place, sit on your right heel.

(d) Place your left elbow forward of your left knee, resting the flat portion of your upper arm on your knee.

(e) Move the rifle butt into the pocket of your right shoulder, pulling the rifle magazine with your right hand and grasping the barrel grip with your left hand.

(f) With your right hand on the rifle magazine, place your right forefinger in the trigger guard of the grenade launcher.

(g) Pull the rifle firmly into your shoulder.

(h) Pull your right elbow in close to your body to help you apply rearward pressure to the weapon. Ensure that your leg completes a solid, three-point base for your position.

(4) ***Kneeling Position*, *Leaf Sight*** (right, Figure 5-2).

(a) For ranges greater than 150 meters, place the butt stock of the weapon under your armpit and grip firmly to prevent the weapon from moving.

(b) Lean your head 45 degrees to the right and place the front sight post of the M16 on the desired range. Raise the butt stock and lower the muzzle to obtain the proper sight alignment and sight picture.

Figure 5-2. Kneeling position, quadrant sight (L) and leaf sight (R).

(5) ***Sitting Position, Open-Legged, Quadrant Sight*** (left, Figure 5-3, page 5-4).

(a) Sit down, breaking your fall with your right hand, and slide your buttocks well to the rear. Face the target half right, and spread your feet wide.

(b) Grasp the rifle magazine with your right hand and the barrel grip with your left hand.

(c) Bend forward from your hips.

(d) Move the butt of the rifle into the pocket of your right shoulder, still holding the rifle magazine with your right hand.

(e) Pull the weapon down slightly with your left hand and pull it to the rear firmly with your right hand.

(6) ***Sitting Position, Open-Legged, Lea f Sight*** (right, Figure 5-3).

(a) For ranges greater than 150 meters, place the butt stock of the weapon under your armpit and grip firmly to prevent the weapon from moving.

(b) Lean your head 45 degrees to the right and place the M16 front sight post on the desired range. Raise the butt and lower the muzzle to obtain the proper sight alignment and sight picture.

Figure 5-3. Sitting position, open-legged, quadrant sight (L) and leaf sight (R).

(7) ***Sitting Position, Cross-Ankle, Quadrant Sight*** (left, Figure 5-4).

(a) Sit facing the target half right.

(b) Extend your legs from your body and cross your left ankle over your right ankle.

(c) Keep both ankles straight.

(d) Grasp the rifle magazine with your right hand and the barrel grip with your left.

(e) Place your left upper arm across your left knee.

(f) Move the butt of the rifle into the pocket of your right shoulder.

(8) ***Sitting Position, Cross-Ankle, Leaf Sight*** (right, Figure 5-4).

(a) For ranges greater than 150 meters, place the butt stock of the weapon under your armpit and grip firmly to prevent the weapon from moving.

(a) Lean your head 45 degrees to the right and place the M16 front sight post on the desired range. Raise the butt and lower the muzzle to obtain the proper sight alignment and sight picture.

Figure 5-4. Sitting position, cross-ankled, quadrant sight (L) and leaf sight (R).

(9) ***Sitting Position, Cross-Legged, Quadrant Sight*** (left, Figure 5-5).

(a) Sit down facing the target half right.

(b) Cross your left leg over your right leg and draw both feet close to your body.

(c) Grasp the rifle magazine with your right hand.

(d) Move the butt of the rifle into the pocket of your right shoulder, and grasp the rifle barrel grip properly with your left hand.

(10) ***Sitting Position, Cross-Legged, Leaf Sight*** (right, Figure 5-5).

(a) For ranges greater than 150 meters, place the butt stock of the weapon under your armpit and grip firmly to prevent the weapon from moving.

(b) Lean your head 45 degrees to the right and place the M16 front sight post on the desired range. Raise the butt and lower the muzzle to obtain the proper sight alignment and sight picture.

Figure 5-5. Sitting position, cross-legged, quadrant sight (L) and leaf sight (R).

(11) ***Squatting Position, Cross-Legged, Quadrant Sight*** (left, Figure 5-6, page 5-6).

(a) Turn half right to the target and, keeping both feet flat on the ground and a comfortable distance apart, squat as low as you can.

(b) Grasp the rifle magazine with your right hand.

(c) Place your left upper arm inside your left knee and the butt of the rifle into the pocket of your right shoulder. Grasp the rifle barrel grip properly.

(d) Lower your right elbow against the inside of your right knee.

(12) ***Squatting Position, Cross-Legged, Leaf Sight*** (right, Figure 5-6, page 5-6).

(a) For ranges greater than 150 meters, place the butt stock of the weapon under your armpit and grip firmly to prevent the weapon from moving.

(b) Lean your head 45 degrees to the right and place the M16 front sight post on the desired range. Raise the butt and lower the muzzle to obtain proper sight alignment and sight picture.

Figure 5-6. Squatting position, quadrant sight (L) and leaf sight (R).

(13) ***Fighting Position, Quadrant Sight*** (left, Figure 5-7).

(a) If possible, use support when firing from a fighting position.

(b) Place your right foot against the rear of the fighting position and lean forward until your chest is against its forward edge.

(c) Grasp the magazine with your right hand.

(d) Place your left elbow on or against solid support.

(e) Use your right hand to position the butt of the rifle in the pocket of your right shoulder. Grasp the rifle barrel grip properly.

(f) Place your right elbow on or against a solid support and relax into a comfortable firing position.

(14) ***Fighting Position, Leaf Sight*** (right, Figure 5-7).

(a) For ranges greater than 150 meters, place the butt stock of the weapon under your armpit and grip firmly to prevent the weapon from moving.

(b) Lean your head 45 degrees to the right and place the M16 front sight post on the desired range. Raise the butt and lower the muzzle to obtain proper sight alignment and sight picture.

Figure 5-7. Fighting position, quadrant sight (L) and leaf sight (R).

NOTE: The weapon must not touch the support.

(15) ***Standing Position, Quadrant Sight*** (left, Figure 5-8).

(a) Face the target while standing with your feet spread a comfortable distance apart.

(b) Grasp the rifle barrel grip with your left hand and the rifle magazine with your right hand.

(c) Place the butt of the stock into your right shoulder so that the sight is level with your eyes.

(d) Hold your right elbow high to form a good pocket for the butt of the stock and to permit a strong rearward pressure with your right hand.

(e) Hold most of the weight of the weapon with your left hand.

(f) Shift your feet until you achieve a natural aiming stance.

(16) ***Standing Position, Leaf Sight*** (right, Figure 5-8).

(a) For ranges greater than 150 meters, place the butt stock of the weapon under your armpit and grip firmly to prevent the weapon from moving.

(b) Lean your head 45 degrees to the right and place the M16 front sight post on the desired range. Raise the butt and lower the muzzle to obtain the proper sight alignment and sight picture.

Figure 5-8. Standing position, quadrant sight (L) and leaf sight (R).

b. **Aiming**. Aiming procedures for every position follow:

(1) ***Aligning Sight***. When using the leaf sight, align it with the front sight post of the M16. When using the quadrant sight, align its rear sight aperture with its front sight post. Picture a horizontal line through the center of the leaf sight or rear sight aperture: the top of the M16's front sight post should touch this line. Picture a vertical line through the center of the leaf sight or rear sight aperture: this line should vertically bisect the front sight post (Figure 5-9, page 5-8).

(2) ***Focusing***. For either sight, focus on the front sight post. A good firing position places your eye directly on line with the center of the leaf sight or rear sight aperture. Your eye's natural ability to center objects in a circle and seek the point of greatest light will help you align the sight correctly.

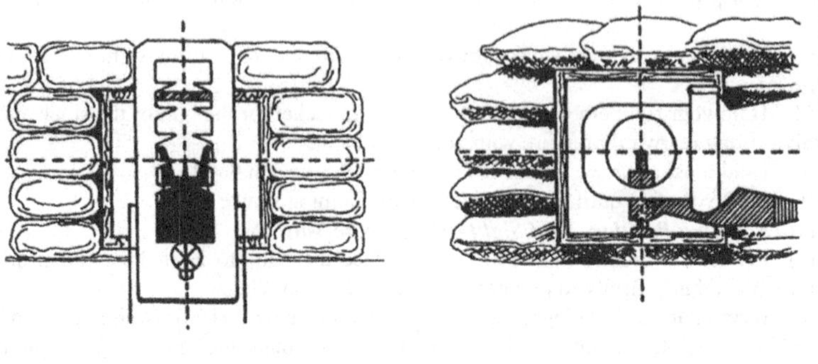

Figure 5-9. Sight pictures for leaf and quadrant sights.

(3) ***Obtaining Sight Picture***. To achieve a correct sight picture, align the front sight post and the leaf sight or rear sight aperture with the target. For area targets, aim where the round's bursting radius will make the round most effective. For point targets, aim at the target's center of mass.

c. **Breathing**. The technique for breathing is the same for every position: Breathe naturally, exhale most of your air, hold your breath, and fire before you become uncomfortable. In combat, just choke off your breath before firing.

d. **Trigger Control**. The technique for trigger control is the same for every position. Place your trigger finger (the index finger of your right hand) so that the trigger is between the first joint and the tip of your finger (not at the extreme end of your finger). Adjust for your hand size and grip. Then, squeeze your trigger finger to the rear without disturbing the lay of the weapon.

5-2. LIMITED VISIBILITY

The fundamentals of marksmanship are almost the same in limited visibility as in normal visibility.

a. **Steady Position**. An M203 with an AN/PVS-4 mounted on it leans to the left. When assuming a steady position, the grenadier must apply more rearward pressure to compensate for the lean and then steady the weapon.

b. **Aiming**. The grenadier sights with the reticle of the AN/PVS-4 rather than with the M203's iron sights. Sighting this way requires him to change position, which breaks his stock weld and makes the weapon seem heavier.

c. **Breathing**. Though breathing itself is affected little by limited visibility, using night vision devices that magnify the field of view increases the effect of weapon movement caused by breathing.

d. **Trigger Control**. This is the same regardless of visibility conditions. The objective is to keep the weapon aligned with the target.

*e. **Night Vision Devices**. The AN/PVS-7 is issued for use with the M203, whereas the AN/PVS-4 is normally issued for use with crew-served weapons. M203 gunners may qualify with either device. In a defensive position, the gunner identifies targets during daylight and constructs aiming or elevation stakes. Because the AN/PVS-7 rear sight must be set to the far setting to sense rounds, the gunner cannot see both the M203 sights and the target at the same time. Therefore, stakes are more important with the AN/PVS-7 than with the AN/PVS-4.

f. **Marked-Sling Method**. The best field-expedient method for firing the M203 grenade launcher in limited visibility is the marked-sling method using only the M16 rifle series (Figure 5-10, page 5-10). To use this method, the grenadier must--

(1) Face the target and kneel on the right knee (if firing right-handed), keeping the left foot pointed toward the target.

(2) Loosen the sling and place the forward foot in the sling.

(3) Place the butt of the stock firmly on the ground.

(4) Using the left hand, grasp firmly the upper barrel grip just below the barrel.

(5) Grasp the receiver group with the right hand.

WARNING

Placing your knee against the buttstock could injure your knee.

(6) Ensure the sling is taut and vertical between the front sling swivel and the boot (Figure 5-11, page 5-10). If not, the rounds will impact at a greater range than desired.

(7) Fire several rounds to determine the desired range.

(8) Mark the sling (where it is held to the ground by the foot) with colored tape, paint, ink, or whatever is available. Mark the position of the buckles (Figure 5-12, page 5-10) so that, if either is moved, the grenadier can return them to their original positions and be assured of constant range accuracy.

(9) If the sling gets wet, it may stretch or shrink, indirectly causing the rounds to impact closer or farther than desired.

*5-3. CHEMICAL, BIOLOGICAL, RADIOLOGICAL, AND NUCLEAR ENVIRONMENTS

The fundamentals of marksmanship remain valid in the CBRN environment, but some modifications may be needed to accommodate the equipment.

a. **Steady Position**. Bulky CBRN wear requires the grenadier to press the stock of the weapon more firmly into his shoulder pocket.

b. **Aiming**. Aiming is affected little by CBRN.

c. **Breathing**. Wearing the protective mask makes breathing more difficult. Grenadiers must try to breathe normally to avoid hyperventilating while firing.

d. **Trigger Control**. All Soldiers must wear rubber gloves.

Figure 5-10. Marked-sling method.

Figure 5-11. Front-sling swivel and front of boot.

Figure 5-12. Marked sling and buckle highlighted.

5-4. FIRE COMMANDS

Standard fire commands are explained to grenadiers and are used during all subsequent gunnery training. Trainers give the appropriate elements before each dry-fire or live-fire exercise. The grenadier performs as directed and repeats each element as it is announced. (Chapter 6 discusses fire commands in detail.)

a. **Alert**. The trainer gives the alert as a fire mission. On hearing this, the grenadier loads the weapon and moves the safety lever to "F."

b. **Direction**. The trainer gives the direction to target.

c. **Description**. The trainer describes the target, for example, "Bunker" or "Machine gun position," and the grenadier lays on the target.

d. **Range**. The trainer gives the (estimated) range to the target, for example, "150."

e. **Method of Fire**. The method of fire for either target is three rounds. On the basic range, grenadiers fire at both point and area targets.

f. **Command to Open Fire**. To open fire, the trainer commands COMMENCE FIRING or AT MY COMMAND. When ready, the grenadier announces "Up" and fires or waits for the command to fire. When all grenadiers are ready, the trainer gives the actual command to fire.

5-5. DRY-FIRE EXERCISES

Dry-fire exercises train grenadiers in the techniques of loading, unloading, immediate action, fundamentals of marksmanship, and sight manipulation. These exercises are conducted with TP or dummy rounds. The trainer gives fire commands as appropriate.

a. **Loading and Unloading Exercise**. This trains the grenadier to operate and clear the weapon proficiently. Loading and unloading procedures (Chapter 3) should be practiced with dummy ammunition.

b. **Immediate Action Exercise**. This exercise is conducted with a dummy round and the basic grenade launcher target.

(1) Load the weapon with a dummy round and aim it at one of the targets on the basic grenade launcher range.

(2) Maintain the sight picture while you pull the trigger to simulate firing.

(3) When you are informed that you have a misfire, apply misfire procedures; then continue to fire (Chapter 4).

c. **Aiming Exercise**. This exercise requires the grenadier to simulate firing a dummy round at a target on the basic grenade launcher range.

(1) Maintain your sight picture throughout the firing cycle.

(2) If, after firing, you note that the sight picture has moved, then you were unsteady when you fired.

(3) After each shot, apply immediate action procedures to extract and eject the dummy cartridges. Then recock the barrel assembly.

d. **Sight Setting and Sight Changing Exercises**. These exercises train the grenadier to operate and adjust both quadrant and leaf sights.

(1) ***Range***. Manipulate the sights to different range settings (quadrant sight, 50 to 400 meters; leaf sight, 50 to 250 meters). To learn to make fine adjustments for elevation, manipulate the sights from the minimum to the maximum setting. When you do not have time to adjust the sights, you may adjust the aiming point instead.

(2) ***Windage***. Depress the rear sight aperture left and right and traverse the windage screw across the entire scale.

e. **Dry-Fire Proficiency (Performance) Exam**. Grenadiers practice the dry-fire tasks until they become proficient in operating the weapon; then they take the dry-fire proficiency exam (Appendix C). This exam emphasizes learning by doing. Before he can progress to live firing, each grenadier must demonstrate skill in every task in the exam.

f. **Remedial Training**. Soldiers who fail to pass the performance exam must attend remedial training, after which they retest. Those who pass can help train those having difficulty.

5-6. SENSING AND ADJUSTMENT OF FIRE

The grenadier determines (senses) where the grenade landed relative to the target and then adjusts elevation and deflection.

a. **Sensing**. As soon as the grenade explodes, determine where it exploded with respect to the target. This is called "sensing" (the impact) and has two aspects: range and deviation. Because the casualty radius of the HE round is 5 meters (5 1/2 yards), determine both range and deviation to the nearest 5 meters.

(1) ***Range***. Sense the range as one of the following:

(a) *Short*. The grenade bursts between you and the target.

(b) *Over*. The grenade bursts beyond the target.

(c) *Target*. The grenade hits any part of the target.

(d) *Range Correct*. The grenade bursts slightly left or right of the target, but at the correct range.

(e) *Doubtful*. The grenade burst left or right of the grenadier, but you cannot sense the range.

(2) ***Deviation***. Announce a deviation sensing as either--

- Right or left of the target, or
- On line with the target.

b. **Adjustment of Fire**. To ensure a second-round hit, adjust your fire by sensing the impact of the round and manipulating the sight.

(1) If time allows, whether using the AN/PVS-4 or AN/PVS-7, adjust the sights; if time is critical, adjust the point of aim instead.

(2) If the grenade lands more than 25 meters over or short of the target, adjust the range quadrant to bring the next grenade on target.

(3) If the grenade explodes less than 25 meters from the target, adjust the point of aim to bring the next grenade on target.

(4) If the launcher is properly zeroed, deviation errors are normally small and easily corrected by adjusting the aiming point. A wind strong enough to move the grenade out of its normal trajectory, however, increases the size of the deviation errors. After observing the effect of the wind on the strike of the grenade, compensate for the effect of the wind by aiming into it. This should help bring the next grenade on target. For example, if the grenade bursts to the left and short of the target, sense the strike of the round relative to the target and then adjust an equivalent distance to the right and over the target to achieve a target hit. Watch the flight of the grenade to the target. This helps determine the effect of the wind on the grenade as it moves toward the target. Evaluating and compensating for the wind before firing increases your chances of achieving a first-round hit.

*5-7. LAYOUT OF GRENADE LAUNCHER RANGE

The grenade launcher range is designed for all grenade launchers. Because Soldiers can qualify on this range in all conditions, it prepares grenadiers for combat situations. The range has four stations (Figure 5-13). Minimum range personnel and their duties are the same for M203 qualification firing as they are for other grenadier firing. These personnel include an OIC, NCOIC, safety officer, ammunition NCO, tower operator, station NCOs, primary trainer, and concurrent training trainers. However, local policy may require more personnel. (Appendix D discusses range safety; TC 25-8 provides a detailed setup and target configuration for this range.)

5-8. DESCRIPTION OF RANGE AND TARGETS

The range has four self-contained stations (Figure 5-13). It is 30 meters wide by 500 meters deep and has a no-HE fire zone out to 165 meters. (Grenadiers can fire HE only on Stations 1, 3, and 4.) Targets should be built from durable materials to reduce downrange target maintenance. Those within each station must be grouped and spaced so that the grenadier may fire on close-range, mid-range, and long-range targets, in that order. The following description of the stations and targets is included to help trainers maintain control during zeroing, practice, and record fire. Figure 5-14 shows the range in meters and feet for combat, point target, training, area target and maximum ranges for the M203.

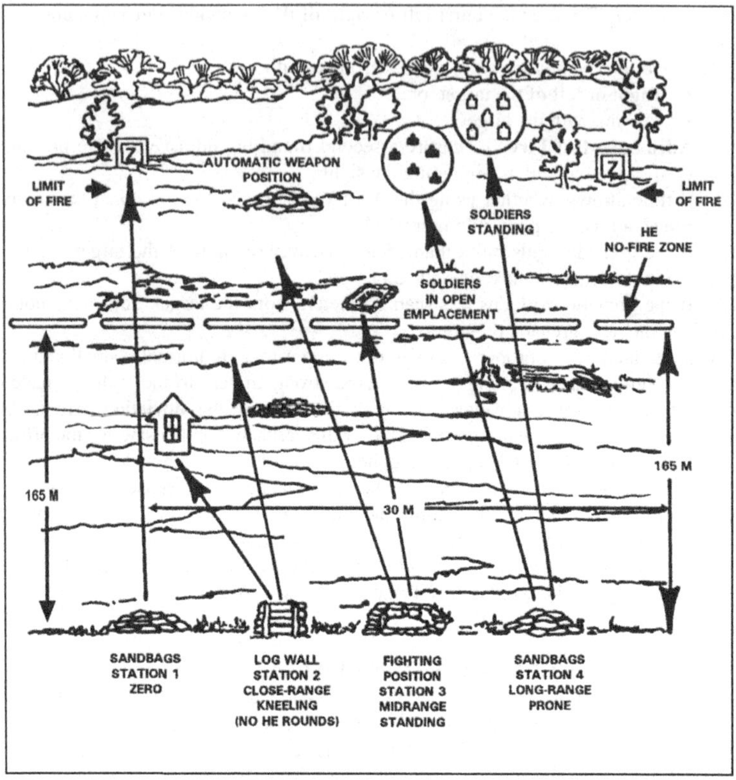

Figure 5-13. Grenade launcher range.

> **WARNING**
>
> **Before firing HE grenades, move to a protected (covered) position. The minimum safe ranges follow:**
>
> **Training 165 meters (541 feet)**
>
> **Combat 31 meters (102 feet)**

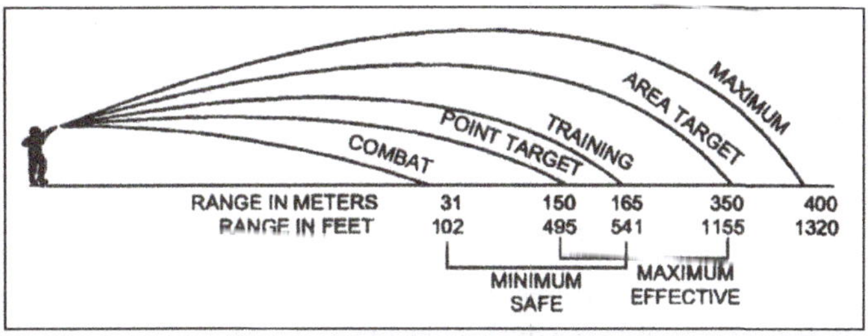

Figure 5-14. Ranges in meters and feet.

NOTE: To simplify the task of finding and destroying duds, trainers must ensure the impact area is free of any vegetation other than short grass.

a. Station 1 consists of a prone fighting position with a log or sandbag support and a zeroing target at 200 meters. The target should be constructed of logs or other suitable material. It must have a surface at least 2 meters high by 2 meters wide (6 feet by 6 feet). The target should be clearly marked with a large "Z" painted in a color that contrasts with the surrounding background and that is visible in different sun or glare conditions.

b. Station 2 consists of an upright log or log wall, a kneeling firing position about 4 feet high, and 2 point-type targets. The targets include a simulated window or door of a building at 90 to 100 meters and a small bunker or fighting position with overhead cover at 125 meters. The targets may be constructed of logs, sandbags, or other suitable material.

c. Station 3 consists of a fighting position and two targets. The targets are a two-person bunker at 150 meters and an automatic weapon position at 175 meters. The bunker represents a point target, and the automatic weapon position represents a target that can be engaged with area-type fire. The targets may be constructed of logs, sandbags, or other suitable material.

d. Station 4 consists of a prone fighting position with a log or sandbag support and two area-type targets (with personnel targets in the open) at 250 and 350 meters. The log or sandbags at the firing position are used for support and cover. The targets are E-type and F-type shaped silhouette and are made out of durable materials.

Section II. BASIC GUNNERY

Basic gunnery allows the grenadier to zero and apply the fundamentals of marksmanship during live-fire exercises in day, night, and CBRN conditions.

5-9. ZEROING OF M203 GRENADE LAUNCHER

A correct zero consists of the elevation and windage sight settings that enable the grenadier to hit the point of aim at a given range with one of the three sighting systems: leaf, quadrant, or night sight. To zero the M203 using either the leaf sight or quadrant sight, the grenadier engages a target at 200 meters. (The M203 is normally zeroed using only the quadrant sight, but may be zeroed with both sights or with only the leaf sight.)

a. **Zeroing the Leaf Sight.** A red mark at 50 meters on the leaf sight reminds the grenadier not to zero at this range.

(1) Select a target at 200 meters.

(2) Place the sight in the upright position.

(3) Place the center mark of the windage scale on the index line on the rear of the sight base.

(4) Loosen the elevation adjustment screw on the leaf sight.

(5) Place the leaf sight's index line on the sight mount's center elevation mark.

(6) Tighten the elevation adjustment screw.

(7) Assume a prone supported firing position.

(8) Load one round of 40-mm HE or TP ammunition.

(9) Use correct sighting and aiming procedures to align the target with the front leaf sight.

(10) Fire a round, sense the impact, and adjust the sight.

(a) *Windage.* Turn the sight windage screw clockwise to move the leaf sight to the left and vice versa. One increment moves round impact 1 1/2 meters at a range of 200 meters.

(b) *Range.* Use a 40-mm cartridge case and turn the elevation adjustment screw to raise the leaf sight (this increases range) or to lower the leaf sight (this decreases range). Moving the leaf sight one increment moves the impact of the round 10 meters at a range of 200 meters.

(11) Fire two more cartridges, readjusting the sight after each. Once a round impacts within 5 meters of the target, the weapon is zeroed.

(12) After you have zeroed the weapon, record the zero data on your scorecard. As soon as you can, transfer the information to a small piece of paper and tape this inside the M16 pistol grip.

b. **Zeroing the Quadrant Sight.**

(1) Select a target at 200 meters.

(2) Ensure that the quadrant sight is correctly mounted on the rifle's carrying handle.

(3) Open the front sight post and rear sight aperture.

(a) Move the front sight post to its highest position and then back 2 1/2 turns.

(b) Depress the rear sight retainer. Slide the rear sight aperture to the left until its white index line aligns with the edge of the sight aperture arm.

(4) Move the sight latch rearward and reposition the quadrant sight arm to zeroing range (200 meters).

(5) Assume a prone supported firing position.

(6) Use correct sighting and aiming procedures to align the target with the front sight post and rear sight aperture.

(7) Load one round of 40-mm HE or TP ammunition.

(8) Fire a round, sense the impact, and adjust the sight.

(a) *Elevation.* Turn the front sight post right (to decrease elevation) or left (to increase elevation). At a range of 200 meters, one full turn equals 5 meters.

(b) *Windage.* Press the sight aperture retainer. Move the rear sight aperture away from the barrel to move the trajectory to the left or vice versa. At a range of 200 meters, one notch on the rear sight aperture equals 1 1/2 meters.

(9) Fire two more cartridges, readjusting the sights after each. If the round lands within 5 meters of the target, the weapon is zeroed.

(10) After you have zeroed the weapon, record the zero data on your scorecard. As soon as you can, transfer this information to a small piece of paper and tape this inside the M16 pistol grip.

5-10. OVERALL QUALIFICATION STANDARDS

DA Form 2946-R (40-mm Grenade Launcher Scorecard) is used for qualification firing. To qualify with an M203, a grenadier must perform to prescribed standards and must score at least 60 of 90 possible points. Each target hit is worth 10 points. Zeroing is not included on the scorecard because the weapon must be zeroed before qualification firing. However, the zero data should already have been entered on the scorecard when the weapon was zeroed. HE familiarization may be included in qualification firing but is not scored. Ratings are awarded based on the point chart shown on the scorecard. A blank scorecard is provided in back of this field manual and must be locally reproduced on 8 1/2- by 11-inch paper. Figure 5-15, page 5-18 shows an example of a completed scorecard.

*5-11. DAY RECORD FIRE

Day record fire gives the grenadier the confidence and experience he needs to progress from dry-fire exercises to record fire. All Soldiers must be prepared to accomplish their missions, even in protective clothing; thus, day record fire includes two CBRN tasks (Tasks 4 and 5). This exercise is conducted on a grenade launcher range IAW Firing Table I (Table 5-1, page 5-20). Before they fire for qualification, grenadiers must first zero their weapons and receive instruction on the objectives, range, targets, and qualification standards. The unit is organized in firing orders based on range constraints. Each firing order consists of two grenadiers, one of whom assists. Grenadiers fire this exercise from the following fighting positions: kneeling supported, mid-range supported, long-range supported, CBRN mid-range point target, and CBRN mid-range area target. For each of these tasks, the grenadier can designate which target he will engage first.

40-MM GRENADE LAUNCHER SCORECARD

For use of this form, see FM 3-22.31; the proponent agency is TRADOC.

EXAMPLE

1. ID CODE *	2. UNIT	3. DATE *(YYYYMMDD)*
20070419BTD01	C CO, 2 29 IN	20070419

4. ZERO LEAF SIGHT	DEFILADE	ELEVATION	5. ZERO QUADRANT SIGHT	DEFILADE	ELEVATION
				L4	R1

6. RECORD FIRE	TASK NO.	TIME (MIN)	TGT	HIT	MISS
DAY	1	2	1	X	
			2		X
	2	2	1	X	
			2	X	
	3	2	1	X	
			2	X	
CBRN	4	2	1	X	
	5	2	1	X	
NIGHT	6	2	1	X	
* Do not use personal information		7. TOTAL POINTS *(Award 10 points for each hit)*		8	

8. RATING SCALE

TOTAL POINTS *(Block 7)*	RATING	
80 - 90	EXPERT	[X]
70 - 75	FIRST CLASS	[ ]
60 - 65	SECOND CLASS	[ ]
0 - 55	UNQUALIFIED	[ ]

9. GRADER'S INITIALS	10. DATE *(YYYYMMDD)*
RM	20070419
11. OIC'S INITIALS	**12. DATE *(YYYYMMDD)***
CS	20070419

DA FORM 2946-R, FEB 2007 EDITION OF NOV 2002 IS OBSOLETE. APD V1.00

Figure 5-15. Example completed DA Form 2946-R, 40-mm Grenade Launcher Scorecard.

WARNING

Before allowing anyone to move between stations, ensure that--

- **All weapons are clear.**
- **Bolts are to the rear.**
- **Barrel assemblies are open.**

If you see an unsafe act, call out "Cease fire" and notify range personnel immediately.

a. **Station 1, Zeroing**. The grenadier zeroes with both quadrant and leaf sights at Station 1.

(1) ***Leaf Sight***. From a prone supported firing position, fire to zero the weapon. This reinforces the experience gained during dry firing and allows practice in loading and firing with the most accurate sensing and adjustments obtainable. If you zero in three rounds, use the other two rounds to confirm the zero. If you cannot zero with five rounds, the trainer must remove you from the firing line for remedial training.

(a) Prepare the sight for zeroing.

(b) Assume a good prone supported firing position.

(c) When you receive the following fire command, repeat each element as it is given:

GRENADIER
FRONT
200 (ZERO PANEL)
ONE ROUND
COMMENCE FIRING

(d) Load one round, obtain the proper sight picture, and announce "Up" to your assistant.

(e) When the tower operator gives the command to commence firing, fire one round at the panel marked "Z."

(f) Sense the impact of the round. If the round did not land within 5 meters of the zero panel, adjust the sights for windage and elevation.

(g) Repeat until a round lands within 5 meters of the zero panel.

(h) Once you have zeroed the weapon, record the zero data on your scorecard. As soon as you can, transfer the information to a small piece of paper and tape this inside the M16 pistol grip.

(2) ***Quadrant Sight***. From a prone supported firing position, fire to zero the weapon. This reinforces the experience gained during dry firing and gives you practice in loading and firing with the most accurate sensing and adjustments you can obtain. The steps for zeroing with the quadrant sight are the same as those for zeroing with the leaf sight.

b. **Station 2, Task 1, Kneeling Supported Position**. (Only TP rounds may be used at this station.)

(1) When you receive the command DESIGNATE THE TARGET, identify the target you intend to engage by announcing WINDOW or BUNKER.

(2) When you receive the command DETERMINE THE RANGE, announce the range to the target.

(3) Load one of the three rounds allotted. Because HE must not be fired at ranges of less than 165 meters on the basic grenade launcher range, use only TP rounds.

(4) When you receive the following fire command, repeat each element as it is given:

GRENADIER
FRONT
3 ROUNDS
90 meters (window) or 100 meters (bunker)
COMMENCE FIRING

(5) Acquire the proper sight picture and announce "Up" to the grader.

(6) Engage the target given in the fire command until you hit it. Fire any remaining rounds at the second target. You need no other fire command. For each round you fire, your assistant announces "Hit" or "Miss."

STATION	TASK	TIME	ROUNDS	TYPE	TARGET(S) AND RANGE(S)
2	1	2 min	3	TP	Window at 90 to 100 meters Bunker at 105 to 125 meters
3	2	2 min	3	HE	Bunker at 170 to 180 meters Automatic weapon at 190 to 200 meters
4	3	2 min	3	HE	Troops in open emplacement at 275 to 300 meters Troops in open at 235 to 350 meters
3	4 CBRN	2 min	3	HE	Bunker at 170 to 180 meters
3	5 CBRN	2 min	3	HE	Automatic weapons at 190 to 200 meters

Table 5-1. Firing Table I, day record fire qualification.

c. **Station 3, Task 2, Midrange Supported Position**.

(1) When you receive the command DESIGNATE THE TARGET, identify the target you intend to engage by announcing "Bunker" or "automatic weapon."

(2) When you receive the command DETERMINE THE RANGE, announce the range to the target.

(3) Load one of the three rounds allotted.

(4) When you receive the following fire command, repeat each element as it is given:

GRENADIER
FRONT
3 ROUNDS
170 METERS (BUNKER) OR 190 METERS (AUTOMATIC WEAPON)
COMMENCE FIRING

(5) Acquire the proper sight picture and announce "Up" to the grader.

(6) Engage the target given in the fire command until you hit it. Fire any remaining rounds at the second target. You need no other fire command. For each round you fire, your assistant announces "Hit" or "Miss."

d. **Station 4, Task 3, Long-Range Supported Position**.

(1) When you receive the command DESIGNATE THE TARGET, identify the target you intend to engage by announcing "Troops in the open emplacement" or "Troops in the open."

(2) When you receive the command DETERMINE THE RANGE, announce the range to the target.

(3) Load one of the three rounds allotted.

(4) When you receive the following fire command, repeat each element as it is given:

GRENADIER
FRONT

3 ROUNDS
300 METERS (TROOPS IN THE OPEN EMPLACEMENT) or
350 METERS TROOPS IN OPEN
COMMENCE FIRING

(5) Acquire the proper sight picture, and announce "Up" to the grader.

(6) When the tower operator gives the command FIRE, engage the target given in the fire command until you hit it. Fire any remaining rounds at the second target. You need no other fire command. Before firing, you must know the procedure to follow in the event of a stoppage. For each round you fire, your assistant announces "Hit" or "Miss."

e. **Station 3, Task 4, Midrange Supported Position, Point Target, CBRN.**

(1) Put on, clear, and check your mask within 9 seconds. Within the next 6 seconds, pull the hood over your head and zip the front of it closed.

(2) Load one of the three rounds allotted.

(3) When you receive the following fire command, repeat each element as it is given:

FIRE MISSION
FRONT
3 ROUNDS
170 METERS (BUNKER)
AT MY COMMAND

(4) Acquire the proper sight picture and announce "Up" to your assistant.

(5) Have your assistant signal the tower operator that you are ready.

(6) When the tower operator gives the command to commence firing, engage the target given in the fire command until you hit it. For each round you fire, your assistant announces "Hit" or "Miss."

f. **Station 3, Task 5, Midrange Supported Position, Area Target, CBRN.**

(1) Load one of the three rounds allotted.

(2) When you receive the following fire command, repeat each element as it is given:

FIRE MISSION
FRONT
3 ROUNDS
190 METERS (AUTOMATIC WEAPON POSITION)
AT MY COMMAND

(3) Acquire the proper sight picture and announce "Up" to your assistant.

(4) Have your assistant signal the tower operator that you are ready.

(5) When the tower operator gives the command to FIRE, engage the target given in the fire command until you hit it. For each round you fire, your assistant announces "Hit" or "Miss."

5-12. DAY RECORD FIRE QUALIFICATION STANDARDS

Before qualification firing, each grenadier must know the tasks, the time and ammunition required, the procedures to follow if a stoppage occurs, the penalties for failure to stop firing when commanded or signaled to do so, and the method used for scoring targets.

a. **Time and Ammunition**. Each grenadier determines the target and its distance before loading any rounds. When the grenadier receives the command to FIRE, the time allotted for that task in Firing Table I begins.

b. **Stoppages**. The grenadier must apply immediate action procedures if a stoppage occurs. If he can reduce the stoppage, he can continue to fire the course. The trainers allow the grenadier an extra 15 seconds for each application of immediate action.

(1) If a stoppage occurs that you cannot reduce by immediate action, raise your hand and announce "Time."

(2) When you say "Time," the assistant trainer notes the time, ensures that a real stoppage exists, and tries to clear the stoppage. If he clears it, you can complete firing. If he cannot clear it, the grader will clear it, and you will be allowed 15 seconds for each round remaining to complete firing.

(3) If you made an error that caused the stoppage, you do not receive extra time, and your score consists only of whatever you had earned when the stoppage occurred.

(4) If the grenade launcher must be replaced, you are allotted 10 rounds to zero a new one; then you may repeat the exercise.

(5) If malfunctions prevent you from finishing the exercise in the time allowed, you can finish it in an "alibi run" after all other grenadiers complete firing.

c. **Penalties**. Five points are deducted from the score of any grenadier who fails to stop firing when the trainer commands or signals to do so. If a grenadier fires at the wrong target, he loses the rounds allotted for the other target, which leaves him only the remainder of his rounds to expend on both targets.

d. **Target Scoring**. The trainer or assistant trainer records scores on DA Form 2946-R. They determine whether each grenade fired is a hit or miss and assign 0 points for a miss or 10 points for a hit. Tasks 1 through 3 each consist of two targets, so the total available for each of these tasks is 20 points. The grenadier may select which of the two targets to engage first. If he scores a hit on the first, the trainer permits him to engage the second. Once he hits both targets, he returns any unexpended rounds to the assistant trainer. Tasks 4 and 5 each consist of firing one target for a total of 10 points each.

(1) ***Window or Door***. To score a hit on a window or door, the grenade must either strike the target or go through the opening in the center of the target.

(2) ***Bunker***. To score a hit on a bunker, the grenade must strike anywhere on the face of the bunker.

(3) ***Automatic Weapon***. To score a hit on an automatic weapon, the grenade must strike within 5 meters of the target.

(4) ***Troops***. To score a hit on troops, the grenade must strike within 5 meters of the target.

5-13. MOUNTING OF AN/PVS-4 (WITHOUT THE RAIL SYSTEM)

The grenadier must mount the AN/PVS-4 to the weapon before he zeroes it, and he must do both before he can qualify with the M203 grenade launcher. To mount the scope, the grenadier must--

a. Remove the quadrant sight.

b. Position the mounting bracket assembly on the left side of the rifle so that the two clamps project through the opening under the handle. Loosen the wing nuts completely (Figure 5-16, page 5-24).

c. Turn the clamp plates so that the pointed ends are in the UP position and are seated against the handle.

d. Tighten the wing nuts clockwise until the mounting bracket is secure against the weapon.

e. Position the sight in the groove on top of the bracket and align the threaded hole in the base of the sight mounting adapter with the lever screw assembly. Tighten the screw clockwise firmly to secure the sight to the bracket.

5-14. ZEROING OF AN/PVS-4 TO M203

After being mounted on the M203, the nightsight must be zeroed to the M16 before it can be zeroed to the M203. FM 3-22.9 provides instructions for doing this. The M16 is then used to zero the nightsight to the M203. The grenade launcher rounds are fired only to confirm the zero. To zero the nightsight to the M203--

a. **Center the Reticle Pattern.** Use the aiming points on the nightsight reticle (Figure 5-17, page 5-24) and the range settings on the mounting bracket. Center the nightsight's reticle pattern within the field of view (FOV). Note that it may not be centered even if it appears to be. To ensure it is, rotate the azimuth control knob either way until it stops. Then, rotate it back the opposite way, counting the number of clicks until it stops again (this may be any number of clicks between 200 and 600). Divide the number of clicks in half and rotate the knob in the original direction by that number of clicks. For example, if the total number of clicks is 500, rotate the knob back 250 clicks in the original direction. Center the elevation using the same procedure with the elevation control knob. The total amount of elevation clicks also varies between 200 and 600.

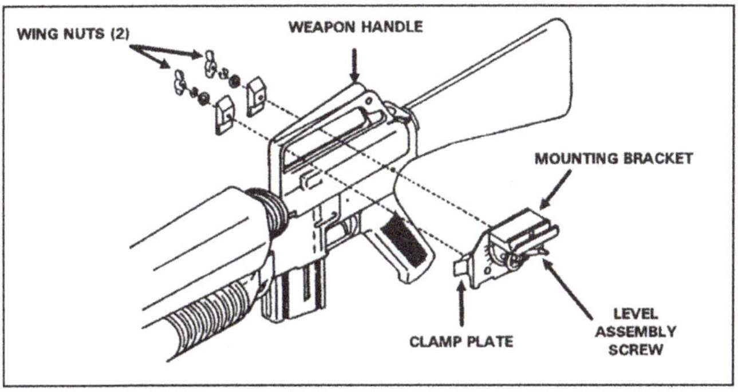

Figure 5-16. Installing the mounting bracket.

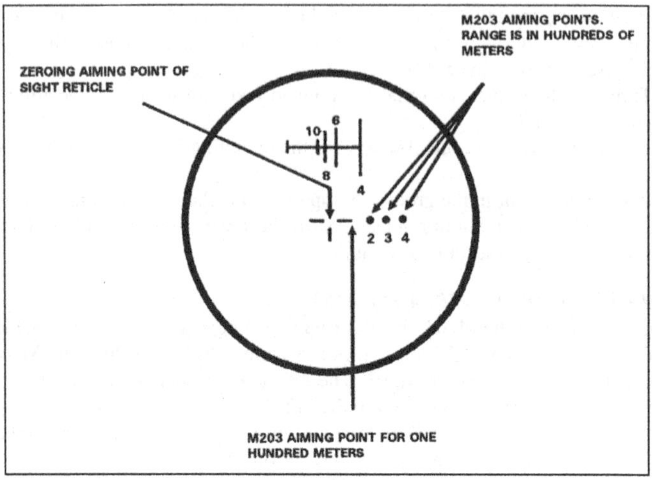

Figure 5-17. Aiming points.

b. **Adjust the Reticle Pattern**. Before adjusting the reticle pattern, the grenadier should fire three 5.56-mm rounds and then retighten the mount wing nuts to securely seat the sight. Once this is done, the grenadier fires at a 10-meter target because hitting and spotting this target is easier than hitting a 25-meter target. This procedure may be performed in daylight using the daylight cover:

(1) Turn the sight on and adjust the reticle intensity to the desired level of illumination.

(2) Place an M16 25-meter target at 10 meters and stabilize the weapon.

(3) Fire a 5.56-mm round at the center of the target and mark the hole the round makes.

(a) If the round misses the entire target, reseat the sight exactly as previously described.

(b) If the round hits the target but not within 20 centimeters (8 inches) of the center, adjust the azimuth and elevation controls to bring the impact point toward the center of the target. Then fire a second round. Continue to fire single rounds and adjust the controls until the rounds strike within the desired distance from the center.

(4) Once the reticle is adjusted, move the 25-meter target out to 25 meters and zero the grenade launcher. Do *not* remove the nightsight from the weapon until you have obtained a zero.

c. **Zero at 25 Meters**. This zero is not recorded. To obtain a 25-meter zero, the grenadier must—

(1) Stabilize the weapon.

(2) Center the reticle's zeroing range aiming point on the target aiming point, which is in the center of the target (Figure 5-18). Fire until you obtain a good three-round shot group. Triangulate and locate the center of the shot group.

NOTE: Even if the nightsight is dismounted and remounted on the same weapon, some changes in its zeroing will occur, so it must be zeroed again.

(3) Turn the azimuth and elevation control knobs to adjust the sight reticle. Move the center of the shot group 9.8 centimeters (3 7/8 inches) below and 4.2 centimeters (1 5/8 inches) to the right of the target aiming point (Figure 5-15). For example, if the shot group is high and to the left of the desired impact point, adjust the elevation down (DN) and the azimuth right (RT). One click of the azimuth or elevation adjustment moves the strike of the round .63 centimeter (1/4 inch) at a range of 25 meters. Two clicks move the reticle about one square on the target.

(4) After adjusting the reticle, assume a stable position. Place the reticle aiming point on the target aiming point and fire three more rounds.

(5) Repeat steps 3 and 4 until the rounds strike within a 3.2 centimeter (1 1/4 inch) circle in the desired location 9.8 centimeters (3 7/8 inches) below and 4.2 centimeters (1 5/8 inches) to the right of the aiming point, or until you have fired 12 rounds, whichever occurs first. If you are unable to zero the AN/PVS-4 after 12 rounds, the trainer must send you to remedial training.

(6) Confirm the zero on the grenade launcher range using a 200-meter target. Place the nightsight into operation and use its reticle, which has two parts. Use the vertical line in the upper part of the reticle to estimate range and the lower part to aim the weapon.

(a) Set the range as estimated on the range indicator of the mounting bracket (Figure 5-16).

(b) Engage the target, placing the aiming point of the sight reticle on the target's center of mass (Figure 5-18). Fire the weapon using all your marksmanship skills. You have confirmed the zero if two of three rounds strike within 5 meters of the target.

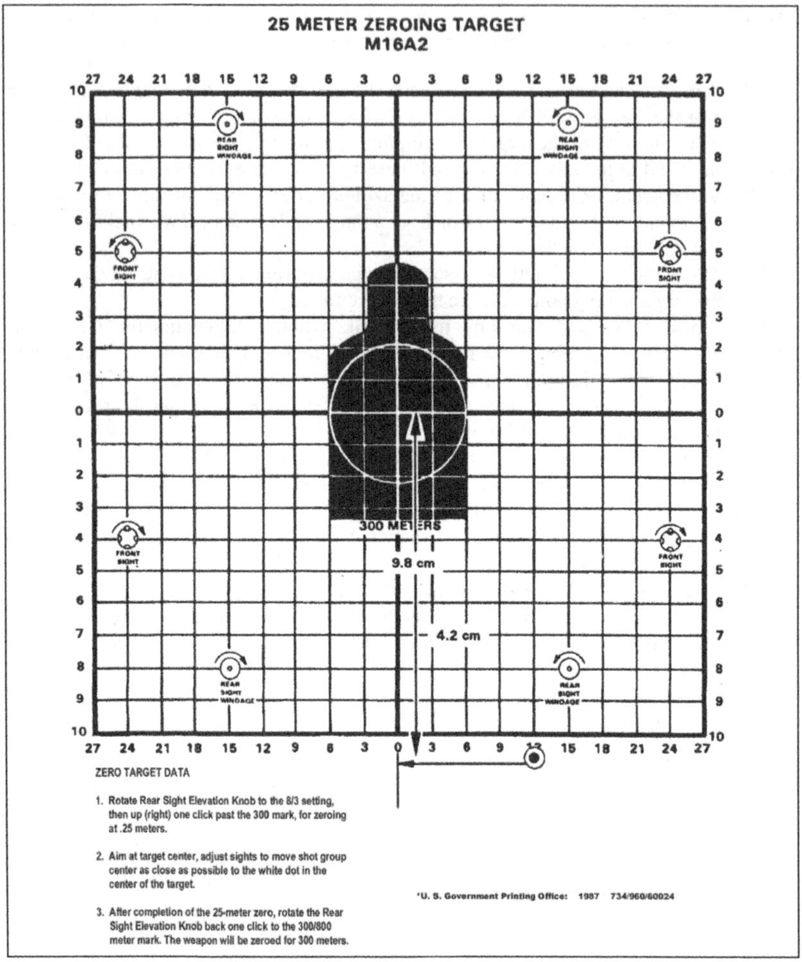

Figure 5-18. Adjustment of rounds.

5-15. NIGHT RECORD FIRE

Night or limited visibility firing trains grenadiers to apply the fundamentals of grenade launcher marksmanship while using the AN/PVS-4 nightsight. This training increases the grenadiers' confidence. Each grenadier learns to zero the M203 to the AN/PVS-4 on the 25-meter range and then fires at one area target on the M203 grenade launcher range. Before night firing, grenadiers receive instruction in its objectives, fundamentals, fire commands, and targets. The unit is organized in firing orders, each consisting of a grenadier and assistant, based on the range constraints. The assistant performs his duties in a manner similar to day record fire. Night record fire consists of one task, *Station 3, Task 6, Mid-Range Fighting Position, Area Target, Night*:

a. Load one of the three rounds allotted.

b. When you receive the following fire command, repeat each element as it is given:

GRENADIER
FRONT
3 ROUNDS
190 METERS (AUTOMATIC WEAPON POSITION)
AT MY COMMAND

c. Acquire the proper sight picture and announce "Up" to the grader.

d. When the grader gives the command FIRE, engage the target given in the fire command until you hit it. Fire any remaining rounds at the second target. You need no other fire command. For each round you fire, your assistant announces "Hit" or "Miss."

5-16. NIGHT RECORD FIRE QUALIFICATION STANDARDS

Before qualification firing, each grenadier must know the task, the time and ammunition required for each, the procedures to follow if a stoppage occurs, the penalties for failure to stop firing when commanded or signaled to do so, and the method used for scoring targets.

a. **Time and Ammunition**. Firing Table II (Table 5-2) provides the night firing task and its time and ammunition requirements.

b. **Stoppages**. The procedure for stoppages is the same as for other qualification firing exercises.

c. **Penalties**. The procedure for penalties is the same as for other qualification firing exercises.

d. **Target Scoring**. The target scoring procedure is the same as for other qualification firing exercises. Record the score in the appropriate block on DA Form 2946-R.

e. **Conditions**. Night record fire trains the grenadier to engage targets between 165 and 250 meters under ideal moonlight conditions.

STATION	TASK	TIME	ROUNDS	TYPE	TARGET(S) AND RANGE(S)
3	3	2 min	3	HE	Automatic weapon at 190 to 200 meters

Table 5-2. Firing Table II, night record fire qualification.

This page intentionally left blank.

(3) ***Linear Targets with Depth***. The grenadier lays on the target's center of mass. He then moves the grenade launcher left and right across the target, selecting successive aiming points at different ranges (Figure 6-9).

Figure 6-9. Linear targets with depth.

* 6-6. DECONTAMINATION

Leaders must try to reduce the penetration of contaminants and lessen exposure to them. Contaminated material is disposed of in accordance with the SOP. Refer to Chapter 2 for decontamination procedures.

Section II. FIRE CONTROL

Fire control includes all leader and Soldier actions in planning, preparing, and applying fire on a target. The leader selects and designates targets, indicating their width and depth or, in the case of targets that are hard to identify, designates the distance from a reference point to the target's center of mass (Figure 6-10, page 6-10). He also designates the midpoint, flanks, or ends of a target unless these locations are obvious to the grenadiers. The grenadiers open fire when ready, adjust and regulate the rate of fire, and shift from one target to another. They cease fire only when the target is neutralized or the leader signals to cease fire.

Figure 6-10. Use of a reference point to identify a target.

6-7. METHODS OF FIRE CONTROL

The noise and confusion of battle may limit the methods of fire control used, so the leader must select the method(s) that will best accomplish the mission.

a. **Oral Commands**. The primary method of fire control is the oral fire command. This method is effective unless noise or distance prevents the grenadier from hearing the leader.

b. **Arm-and-Hand Signals**. This method of fire control is effective only if the grenadiers know the standard arm-and-hand signals and can see the leader. The leader gets the grenadier's attention, then points to the target. When the grenadier returns the "ready" signal, the leader commands the grenadier to fire.

c. **Prearranged Signals**. This method of fire control can include visual or sound signals such as those that can be produced by a whistle, pyrotechnics, or casualty-producing device. The SOP must define the signals to be used, and all squad members must understand them. If the leader wants to shift fire at a certain time, he gives a prearranged signal such as smoke or pyrotechnics. When they see this signal, grenadiers shift their fire to a prearranged point.

d. **Personal Contact**. This method of fire control is the one most frequently used by leaders of small units. Many situations require the leader to move to individual Soldiers to issue orders. If so, he must use cover and concealment to avoid disclosing their positions. Once there, he gets the grenadier's attention, points out the new target, and commands FIRE.

e. **Standing Operating Procedure**. This method of fire control refers to actions executed without command. The SOP defines these actions and the events that initiate them. Using an SOP simplifies the leader's job of fire control.

f. **Range Cards**. This method of fire control requires the leader to ensure all range cards are current and accurate. He designates dead spaces, specific targets, no-fire zones, and restricted fire areas. The key to this method is the disciplined grenadier who pays attention to detail and can understand the areas the squad leader wants covered by fire. Figure 6-11 shows an example of an M203 range card.

*GLOSSARY

AAR	after-action review
ammo	ammunition
C	Centigrade
CBRN	chemical, biological, radiological, and nuclear
CCC	Captain's Career Course
CLP	cleaner, lubricant, preservative
cm	centimeter(s)
CMF	career management field
CS	a chemical agent (tear gas)
DA	Department of the Army
DODAC	Department of Defense Ammunition Code
EST	engagement skills trainer
F	Fahrenheit
FM	field manual
FOV	field of view
FPL	final protective line
fps	feet per second
FTX	field training exercise
HE	high explosive
HEDP	high-explosive dual-purpose
hr	hour(s)
IAW	in accordance with
IET	initial entry training
KD	known-distance (range)
kg	kilogram(s)
kps	kilometers per second
LFX	live-fire exercise
LTA	local training area
METL	mission-essential task list
m	meter(s)
min	minute(s)
mm	millimeter(s)
MOPP	mission-oriented protective posture
mph	mile(s) per hour

mps	meter(s) per second
MTA	major training area
MTP	mission training plan
NCO	noncommissioned officer
NCOES	Noncommissioned Officer Education System
NCOIC	noncommissioned officer in charge
NSN	national stock number
NVD	night vision device
OBC	Officer Basic Course
OIC	officer in charge (of)
OGIVE	head of the round
PD	point-detonating
PDF	principal direction of fire
PMT	preparatory marksmanship training
psi	pounds per square inch
RETS	remoted target system
rpm	revolution(s) per minute
S2	Intelligence Officer
S3	Operations and Training Officer
S4	Supply Officer
SEE	small emplacement excavator
SM	soldier's manual
SMCT	soldier's manual of common tasks
SOP	standing operating procedure
SSN	social security number
STP	soldier training publication
STX	situational training exercise
TP	training practice
TRADOC	Army Training and Doctrine Command
TRC	training readiness condition
UO	urban operations

*REFERENCES

SOURCES USED

These are the sources quoted or paraphrased in this publication. All are US Army publications unless stated otherwise.

FM 3-3	Chemical and Biological Contamination Avoidance. 16 November 1992, with Change 1, 29 September 1994.
FM 3-11.4	Multiservice TTP for NBC Protection. 2 June 2003.
FM 3-11.5	Multiservice TTP for Chemical, Biological, Radiological, and Nuclear Decontamination. 4 April 2006.
FM 21-60	Visual Signals. 30 September 1987.
FM 21-75	Combat Skills of the Soldier. 3 August 1984.
FM 3-22.9	Rifle Marksmanship, M16A1, M16A2/3, M16A4, and M4 Carbine. 24 April 2003, with Changes 1 through 3, 22 January 2004 through 28 April 2005.
STP 7-11B1-SM-TG	Soldier's Manual and Trainer's Guide, MOS 11B, Infantry, Skill Level 1. 28 May 2004.
STP 7-11B24-SM-TG	Soldier's Manual and Trainer's Guide, MOS 11B, Infantry, Skill Levels 2, 3, and 4. 28 May 2004.
STP 21-1-SMCT	Soldier's Manual of Common Tasks, Skill Level 1. 11 October 2005.
TC 25-8	Training Ranges, 25 February 1992, with Change 1, 5 April 2005.
TM 9-1010-221-10	Operator's Manual for 40-mm Grenade Launcher, M203. 1 August 2001.
TM 9-1010-221-23&P	Unit and Direct Support Maintenance Manual, Including Repair Parts and Special Tools List, for Launcher, Grenade, 40-mm, M203. 12 January 1993.
TM 43-0001-28	Army Ammunition Data Sheets for Artillery Ammunition: Guns, Howitzers, Mortars, Recoilless Rifles, Grenade Launchers, and Artillery Fuzes. 28 April 1994 with Changes 1 through 13, 30 May 1995 through 31 March 2005.

TM 750-244-7 — Procedures for Destruction of Equipment In Federal Supply Classifications 1000, 1005, 1010, 1015, 1020, 1025, 1030, 1055, 1090, and 1095 To Prevent Enemy Use. 18 June 1970, with Change 1, 29 October 1970.

DOCUMENTS NEEDED

These documents must be available to the intended users of this publication.

AR 385-63 — Policies and Procedures for Firing Ammunition for Training, Target Practice, and Combat. 19 May 2003.

DA Form 2946-R — 40-mm Grenade Launcher Scorecard.

DA Pam 350-38 — Standards in Weapons Training. April 2006.

TRADOC Reg 350-6 — Enlisted Initial Entry Training (IET) Policies and Administration. 30 December 2005.

WEB SITES

Army Knowledge Online	https://akocomm.us.army.mil/usapa/doctrine/index.html
Army Publishing Directorate (for forms)	http://www.apd.army.mil/usapa_PUB_formrange_f.asp
Reimer Digital Library	http://www.train.army.mil

*INDEX

40-MM GRENADE LAUNCHER SCORECARD

For use of this form, see FM 3-22.31; the proponent agency is TRADOC.

1. ID CODE *	2. UNIT	3. DATE *(YYYYMMDD)*

4. ZERO LEAF SIGHT	DEFILADE	ELEVATION	5. ZERO QUADRANT SIGHT	DEFILADE	ELEVATION

6. RECORD FIRE	TASK NO.	TIME (MIN)	TGT	HIT	MISS
DAY	1	2	1		
			2		
	2	2	1		
			2		
	3	2	1		
			2		
CBRN	4	2	1		
	5	2	1		
NIGHT	6	2	1		
* Do not use personal information		7. TOTAL POINTS *(Award 10 points for each hit)*			

8. RATING SCALE

TOTAL POINTS *(Block 7)*	RATING	
80 - 90	EXPERT	☐
70 - 75	FIRST CLASS	☐
60 - 65	SECOND CLASS	☐
0 - 55	UNQUALIFIED	☐

9. GRADER'S INITIALS	10. DATE *(YYYYMMDD)*
11. OIC'S INITIALS	**12. DATE *(YYYYMMDD)***

DA FORM 2946-R, FEB 2007 EDITION OF NOV 2002 IS OBSOLETE. APD V1.0

Made in the USA
Monee, IL
04 November 2026

67386435R00046